PLANT SCIENCE RESEARCH AND PRACTICES

SUNFLOWERS

CULTIVATION, USES AND ECOLOGICAL SIGNIFICANCE

PLANT SCIENCE RESEARCH AND PRACTICES

Additional books and e-books in this series can be found on Nova's website under the Series tab.

PLANT SCIENCE RESEARCH AND PRACTICES

SUNFLOWERS

CULTIVATION, USES AND ECOLOGICAL SIGNIFICANCE

ÉRICO DE SÁ PETIT LOBÃO
EDITOR

Library of Congress Cataloging-in-Publication Data

Names: De Sá Petit Lobão, Érico, editor.
Title: Sunflowers : cultivation, uses and ecological significance / [Érico de Sá Petit Lobão].
Other titles: Plant science research and practices.
Description: [Hauppauge], New York : Nova Science Publishers, [2019] | Series: Plant science research and practices | Includes bibliographical references and index. |
Identifiers: LCCN 2019059680 (print) | LCCN 2019059681 (ebook) | ISBN 9781536171952 (paperback) | ISBN 9781536171969 (adobe pdf)
Subjects: LCSH: Sunflowers. | Sunflowers--Utilization. | Sunflowers--Brazil.
Classification: LCC SB413.S88 S862 2019 (print) | LCC SB413.S88 (ebook) | DDC 635.9/3399--dc23
LC record available at https://lccn.loc.gov/2019059680
LC ebook record available at https://lccn.loc.gov/2019059681

Published by Nova Science Publishers, Inc. † New York

CONTENTS

PREFACE

Sunflower: Cultivation, Uses and Ecological Significance is a work that brings together renowned researchers from four different countries with experience and scientific competence in the research of Hellianthus annus L. The book consists of six chapters that present information referenced in years of research on various topics such as: a short sunflower review; an organic fertilization revision; the stages of sunflower development; the sunflower under water stress conditions; the sunflower genetic variability; and the development of composite materials.

In each section of this book the authors have provided a little feedback on their extensive work from years of research devoted to the knowledge, domestication and improvement of the use of the species. There are a huge range of sunflower varieties currently being used by modern civilization for different purposes and applications. It is an extremely versatile species that can deliver extraordinary results in different sectors, whether as a highly healthy food in the food industry or as a strategic oilseed for the bioenergy and biofuels segment. It would be impossible to gather all the available information in a single work, however, this small work aims to contribute to the global collection about the species and guide future works that are still needed for the best use of this incredible plant that is the sunflower.

Chapter 1 - Long pre-historical part of sunflower evolution started in the eastern part of the North American continent. As Charles B. Heiser wrote in his prominent paper "The sunflower among the North American Indians"

(1951) "The cultivated sunflower (*Helianthus annuus* var. macrocarpus) was rather widely grown by the North American Indians", and they also used wild sunflower as food. Then the sunflower was brought to Europe by the Spanish. They usually used it as a very unusual ornamental plant. Sunflower was spread throughout Europe. Once upon a time, sunflower came to Russia. This event was not recorded properly and leaves room for discussions. The most popular version connects sunflower and Russian tsar Peter the Great. He introduced many ideas, mechanisms, things and crops into Russia from Europe. It is possible that sunflower was among them. The second version is more practical – German colonists brought it to the Volga region of Russia in the end of XVII or in the very beginning of XVIII century. Taste of sunflower seeds were recovered quickly and sunflower became a very common plant for vegetable gardens in Russia and Ukraine. Custom to eat roasted sunflower seeds spread everywhere. Being just a crop for vegetable gardens, sunflower increased the acreage quickly and became a rather important item for the trade. Some period later a peasant Dmitriy I. Bokarev started to produce oil from the sunflower seeds. It happened in former Alekseev's sloboda (Alekseyevka town of Belgorod region now) in the 1829, and the first sunflower oil-productive plant based on the steam energy began to work already in the 1865. Sunflower as a crop took attention of some researchers and breeders. From the 1910 scientific breeding starts simultaneously in two places – Saratov (E.M. Plachek) and Krasnodar (V.S. Pustovoit). Until that time, people developed several local varieties widely marketed and spread. Pustovoit produced modern oil-type sunflower varieties with high oil content (up to 52% instead of 30-32% before) in the seeds. It allowed sunflower become a world-wide important oil crop. Next important steps were made by Leclercq in 1969 (CMS-pet discovery) and Kinman in 1970 (detection of fertility restorer gene for this cytoplasm). After it F_1 sunflower hybrids with uniformity and high yield potential were developed and replaced OP-varieties on the fields throughout the world.

Chapter 2 - In Brazil, research on sunflower (*Hellianthus annnus* L.) was initiated by the Campinas Agronomic Institute (IAC) of the State of São Paulo, whose first records date from 1932. However, the interest and increase of sunflower cultivation in Brazil occurred mainly due to research and technological development in the sector in the 90s. In addition, the numerous forms of use, the emergence of industries and the need for farmers to diversify their crops, whether in the form of oil for human consumption or for the production of biodiesel, in animal feed through bird seed, bran, pie or silage for ruminants and monogastric, have contributed to increase research and production, as well as for the establishment of its production chain in the different regions of the country. Besides the great challenges for sunflower production and use, nitrogen deficiency has been identified as the most frequent nutritional disorder of the plant causing growth problems and plant stem breakage. Organic fertilization brings physical, chemical and biological benefits. When it comes to organic fertilizers, animal manures are the most important, whether for their composition, relative availability or application benefits.

Chapter 3 - Schneiter and Miller (1981) offered description of sunflower growth stages highlighting several vegetative stages (V1, V2... etc) and reproductive Stages (R1, R2...). The classification was based on external morphological features. The selection of several reproductive stages (R5.1, R5.2 etc) is based on the percentage ratio of blossoming flowers and does not take into account the development of male and female reproductive structures. Toderich (1988) investigated the correlations between male and female reproductive structures development. She identified 8 stages in the development of a tubular flower and described each in details.

Using the authors' own data on the development of male and female reproductive structures in sunflower, the authors propose to combine and improve their classifications. The authors match the number of flowering circles in the head with changes in the reproductive structures - the phases of pollen formation and ovule development. Thus, the external morphological characteristics of the head (inflorescences), as a number of flowering circle, are associated with the peculiarities of the development of the internal structures of the flower. In the authors' version of the classification, a simple and understandable division into the flowering stage of the head with a description of the state of the reproductive organs of the flower and its photographs is proposed.

Chapter 4 - Sunflower is one of the economically important crops in the world. It is largely cultivated for biofuel production, represents a relevant energy source in human diet and is widely used as ornamental plant. However, the water stress is a limiting factor in sunflower cultivation, leading to an emphasis on developing innovative technologies that make it possible to maximize the water use efficiency, as well as to improve irrigation and management strategies. In this context, Factorial Discriminant Analysis (FDA) and Partial Least Squares Discriminant Analysis (PLS-DA) were used to classify intact sunflower plants under three water stress levels (without, moderate, and severe) based on Vis/NIR spectral reflectance measurements (500 - 1000 nm range). The raw and pre-treated (detrend, first and second derivatives) spectral signatures were centered and normalized before data processing. Results showed a satisfying performance when using both multivariate methods with raw and pre-treated spectra. But, PLS-DA using the first derivative pre-treatment performed slight better for classifying the sunflower according to different water stress conditions, reaching overall correct classification of 80.67% when considering the external validation dataset. The lower accuracy was verified for the FDA with second derivative, resulting in an overall correct classification of 69.05%. The Vis/NIR spectroscopy combined with chemometrics is a promising tool for non-destructive and fast identification of sunflower water stress and can be used into processing lines, enabling large-scale individual analysis and real-time decision making.

Chapter 5 - The spectrum and frequency of morphological and physiological mutations after treatment of mature and immature seeds as well as immature embryos of different age (9-11 and 14-16 days old) with chemical mutagen ethyl methanesulfonate (EMS) were studied. Mature and immature (20 days old) seeds were soaked in the mutagen solution of 0.1 and 0.01% concentration. Immature embryos were treated with 0.02% of EMS. Immature seeds and embryos were later grown in vitro on the modified MS medium. It has been proved that the use of immature embryos for mutagenic treatment, as opposed to mature seeds, resulted in wider genetic variability. That was manifested in increasing the frequency of mutations up to two-fold and broadening the mutation spectrum. Generation of some specific mutations was greatly influenced by the age of the embryos. Thus, mutations that cause the disruption of male and female generative structures were observed when treating the 14-16 days old embryos only. Agronomic importance of the selected mutant accessions was demonstrated. The emphasis was made on the use of morphological mutants as a source of the marker traits. Genetic control of some mutant traits was revealed.

Chapter 6 - This chapter discusses the use of sunflower oil as an epoxy resin in biocomposite material. Sunflower oil is characterized by a great proportion of unsaturated fatty acids (oleic, linoleic and linolenic) which confers a high polymerization capacity. For this, sunflower oil was chemically modified by epoxidation and acrylation and then copolymerized to obtain epoxy resin. This new class of bioderived epoxy resins shows a good mechanical and thermal properties and higher resistance to moisture absorption.

Epoxy resin was used as a matrix and alfa fiber as reinforcement to prepare biocomposites materials which were characterized in terms of mechanical and thermal properties.

A biodegradation study was carried out for the synthesized biocomposites by measuring the CO_2 level obtained by means of a laboratory respirometry test and by the measurement of the mass loss in a solid medium (burial in the soil) during one year.

The results showed that biodegradation of the biocomposites occurred.

In: Sunflowers
Editor: Érico de Sá Petit Lobão
ISBN: 978-1-53617-195-2

Chapter 1

SHORT SUNFLOWER HISTORY REVIEW

Sergey Gontcharov*
Kuban State Agrarian University, Krasnodar, Russia

ABSTRACT

Long pre-historical part of sunflower evolution started in the eastern part of the North American continent. As Charles B. Heiser wrote in his prominent paper "The sunflower among the North American Indians" (1951) "The cultivated sunflower (*Helianthus annuus* var. macrocarpus) was rather widely grown by the North American Indians", and they also used wild sunflower as food. Then the sunflower was brought to Europe by the Spanish. They usually used it as a very unusual ornamental plant. Sunflower was spread throughout Europe. Once upon a time, sunflower came to Russia. This event was not recorded properly and leaves room for discussions. The most popular version connects sunflower and Russian tsar Peter the Great. He introduced many ideas, mechanisms, things and crops into Russia from Europe. It is possible that sunflower was among them. The second version is more practical – German colonists brought it to the Volga region of Russia in the end of XVII or in the very beginning of XVIII century. Taste of sunflower seeds were recovered quickly and sunflower became a very common plant for vegetable gardens in Russia and Ukraine. Custom to eat roasted sunflower seeds spread everywhere. Being just a

* Corresponding Author's Email: serggontchar@hotmail.com.

crop for vegetable gardens, sunflower increased the acreage quickly and became a rather important item for the trade. Some period later a peasant Dmitriy I. Bokarev started to produce oil from the sunflower seeds. It happened in former Alekseev's sloboda (Alekseyevka town of Belgorod region now) in the 1829, and the first sunflower oil-productive plant based on the steam energy began to work already in the 1865. Sunflower as a crop took attention of some researchers and breeders. From the 1910 scientific breeding starts simultaneously in two places – Saratov (E.M. Plachek) and Krasnodar (V.S. Pustovoit). Until that time, people developed several local varieties widely marketed and spread. Pustovoit produced modern oil-type sunflower varieties with high oil content (up to 52% instead of 30-32% before) in the seeds. It allowed sunflower become a world-wide important oil crop. Next important steps were made by Leclercq in 1969 (CMS-pet discovery) and Kinman in 1970 (detection of fertility restorer gene for this cytoplasm). After it F_1 sunflower hybrids with uniformity and high yield potential were developed and replaced OP-varieties on the fields throughout the world.

The cultivated sunflower (*Helianthus annuus* L.) has a long and dramatic history. First, it belongs to the *Helianthus* genus, which is native to North America. Genera *Phoebantus*, *Viguera* and *Tithonia* considered being the closest relatives of *Helianthus* which includes 67 species (Heiser et al. 1969). Later 16 species native to South America were removed from the *Helianthus* to other genera and now there are 51 different species of sunflowers (Shilling and Heiser 1981; Jan and Seiler 2007), 14 species are annual and 37 – perennial ones. All of them inhabit various localities mostly arid ones, including deserts. Annual species are diploids with chromosome number 2n = 34 (Table 1), among the perennial ones (Table 2) there are diploid, tetraploid (2n = 68) and hexaploid species (2n = 102) (Jan and Seiler 2007). Later, a new perennial species, *Helianthus winteri*, was discovered in California (Stebbins et al. 2013). Now many species of *Helianthus* genera are used as a genetic resource for sunflower breeding.

Now you can see many sunflower species practically throughout the world as field crops (*H. annuus* and *H. tuberosus* – Jerusalem artichoke), ornamental plant species and weeds (Fernandez-Martinez et al., 2009). Breeders, farmers, traders and other human beings helped sunflowers to conquer the world.

The sunflower (*H. annuus*) still grows wild in the USA, Canada, and Mexico (Fick and Miller 1997). Archaeological evidence reveals the use of sunflower (wild and domesticated) among American Indians (Heiser 1955). Semelczi-Kovacs (1975) considers cultivation of sunflower began in Arizona and New Mexico about 3000 BC. Much of the evidence indicates that a type of monocephalic, domesticated sunflower existed in the prehistoric North American Indian culture (Putt 1997). Whiting (1939) cited opinion of V.N. Jones that sunflower may have been domesticated before corn (*Zea mays* L.). Smith (2006) defined eastern North America as a place of sunflower domestication. Here more than 4800 years ago Native Americans started sunflower cultivation. Gathering wild sunflower seeds as a food obviously began earlier. Even later they used them, as Charles B. Heiser wrote in his prominent paper "The sunflower among the North American Indians" (1951) "The cultivated sunflower (*Helianthus annuus* var. macrocarpus) was rather widely grown by the North American Indians" and they also used wild sunflowers as a food. There are some other opinions about the place of sunflower domestication, Mexico for example, but modern investigations based on the molecular markers approach do not support such ideas (Lentz et al. 2001; Smith 2006).

Interesting that sunflower seed size was the most important trait in the domestication process as molecular analyses revealed (Wills and Burke 2007; Burke et al. 2009).

Then the sunflower was brought in Europe by the Spanish. They usually used it as a very incredible ornamental plant. The earliest records were of sunflower seed obtained by a Spanish expedition in 1510 was sown in a botanic garden in Madrid (Zukovsky 1950). Many of researchers consider Dodonaeus as a person who first published description of the sunflower in 1568 (Heiser 1951; Gundaev 1971; Semelczi-Kovacs 1975). After the introduction to Spain, the sunflower plant moved to Italy and France (Putt 1997). Little by little sunflower was spread throughout Europe. In 1716 English patent No. 408 was granted to Arthur Bunyan for inventing a method for using sunflower seed as a source of oil. Sunflower oil was suggested to use for the industrial purpose (Putt 1997).

Table 1. Infrageneric classification of annual *Helianthus* species (n = 17) (Schilling and Heiser 1981; Jan and Seiler 2007)

Section	Species
Helianthus	*H. annuus* L.
	H. anomalus S.F. Blake
	H. argophyllus Torr.& A. Gray
	H. bolanderi A. Gray
	H. debilis Nutt.
	subsp. *Debilis*
	subsp. *cucumerifolius* (Torr. & A. Gray) Heiser
	subsp. *silvestris* Heiser
	subsp. *tardiflorus* Heiser
	subsp. *vestitus* (E. Watson) Heiser
	H. deserticola Heiser
	H. exilis A. Gray
	H. neglectus Heiser
	H. niveus (Benth.) Brandegee
	subsp. *canescens* (A. Gray) Heiser
	subsp. *Niveus*
	subsp. *tephrocles* (A. Gray) Heiser
	H. paradoxus Heiser
	H. petiolaris Nutt.
	subsp. *fallax* Heiser
	subsp. *Petiolaris*
	H. praecox Engelm. & A. Gray
	subsp. *hirtus* (Heiser) Heiser
	subsp. *Praecox*
	subsp. *runyonii* (Heiser) Heiser
Agrestes	*H. agrestis* Pollard
Porteri	*H. porteri (A.* Gray) Pruski

Once upon a time, sunflower came to Russia. This event was not recorded properly and leaves room for discussions. The most popular version connects sunflower and Russian tsar Peter the Great (Zukovsky 1950). He introduced many ideas, mechanisms, things and crops into Russia from Europe. It is possible that sunflower was among them, but it has not yet been proven by the historians. The second version is more practical – German colonists brought it to the Volga region of Russia in the end of XVII or in the very beginning of XVIII century. Taste of sunflower seeds were recovered quickly and sunflower became a very common plant for vegetable gardens in Russia and Ukraine. Custom to eat roasted sunflower seeds spread everywhere.

Table 2. Infrageneric classification of perennial *Helianthus* species (Schilling and Heiser 1981; Jan and Seiler 2007)

Section	Series	Species	Chromosome number (n)
1	2	3	4
Ciliares	*Ciliares*	*H. arizonensis* R.C. Jacks.	17
		H. ciliaris DC.	34, 51
		H. laciniatus A. Gray	17
Ciliares	*Pumili*	*H. cusickii* A. Gray	17
		H. gracilenlus A. Gray	17
		H. pumilus Nutt.	17
Atrorubens	*Coronasolis*	*H. californicus* DC.	51
		H. decapetalus L.	17, 34
		H. divaricatus L.	17
		H. eggertii Small	51
		H. giganteus L.	17
		H. grosseserratus M. Martens	17
		H. hirsutus Raf.	34
		H. maximiliani Schrad.	17
		H. mollis Lam.	17
		H. nuttallii Torr. & A. Gray	
		subsp. *Nuttallii*	17
		subsp. *parishii* (A. Gray) Heiser	17
		subsp. *rydbergii* (W\iion) R. Long	17
		H. resinosus Small	51
		H. salicifolius A. Dietr.	17
		H. schweinitzii Torr. & A. Gray	51
		H. strumosus L.	34, 51
		H. tuberosus L.	51
Atrorubens	*Microcephali*	*H. glaucophyllus* D.M. Sm.	17
		H. laevigatus Torr. & A. Gray	34
		H. microcephalus Torr. & A. Gray	17
		H. smithii Heiser	17, 34
Atrorubens	*Atrorubentes*	*H. alrorubens* L. *H. occidentalis* Riddell	17
		subsp. *occidentalis*	17
		subsp. *plantagineus* (Torr. & A. Gray) Heiser	17
		H. pauciflorus Nutt.	
		subsp. *pauciflorus*	51
		subsp. *subrhomboideus* (Rydb.) 0. Spring & E.E. Schill.	51
		H. silphioides Nutt.	17
Atrorubens	*Angustifolii*	*H. angustifolius* L.	17
		H. carnosus Small	17
		H. floridanus A. Gray ex Chapm.	17
		H. heterophyllus Nutt.	17
		H. longifotius Pursh	17
		H. radula (Pursh) Torr. & A. Gray	17
		H. simulons E. Watson	17
		H. verticillatus Small	17

Being just a crop for vegetable gardens, sunflower increased the acreage quickly and became a rather important item for the trade. Some period later a peasant Dmitry I. Bokarev started to produce oil from the sunflower seeds. It happened in former Alekseev's sloboda (Alekseyevka town of Belgorod region now) in the 1829, and the first sunflower oil-productive plant based on the steam energy began to work already in the 1865. More other sunflower stems were an important source to produce potash for export into Europe in the end of XIX century. Large numbers of local cultivars were available by the 1880s, some of which had improved resistance to broomrape (*Orobanche cumana* Wallr.) and the sunflower moth (*Homoeosoma nebulella* Hb.) (Pustovoit 1964).

By 1880, the crop occupied 150000 ha in Russia (Semelczi-Kovacs 1975). From this time to present, sunflower is the main oil crop in Russia. Sunflower as a crop took attention of some researchers and breeders. From the 1910 scientific breeding starts simultaneously in two places – Saratov (Plachek) and Krasnodar (Pustovoit). Until that time, people developed several local varieties widely marketed and spread. Among them were oil-type and confectionary sunflower varieties. The first type had a small, well-filled, round seed with a thin hull with the oil content ranging between 200 and 300 g/kg. The second one was for direct human consumption with a large, long seed with thick, heavy hull and oil content ranging from only 150 to 200 g/kg (Putt 1997). Primarily main breeding purposes were to reduce vegetation period, improve kernel yield and quality.

People began cultivating sunflower as an oil crop in Hungary practically at the same time as in Russia. There are some records about sunflower oil production at least in 1812-1814. By the end of the XIX century, sunflower was the primary oil crop in Hungary and was exported. Romania, Bulgaria and France also had significant sunflower acreage (Putt 1997).

Sunflower acreage increased and sunflower was sown on 3.14 million ha in the Soviet Union in 1937 (over 80% from total 3.87 million ha in the entire world). Among other sunflower producing countries were Bulgaria (about 180 thousand ha), Hungary (177741 ha), Romania (199995 ha) and Argentina (123930 ha) (Morozov 1947).

Argentina acquired a sunflower from European immigrants and more or less significant commercial production started in the 1920s. The first recorded data of the sunflower area planted in Argentina was 84000 ha in 1934 (Severin 1939). Now Argentina is one of the main producers of sunflower.

North America being the native land for sunflower did not consider it as a crop for a rather long time. It was reintroduced from Europe to the USA and Canada partly by immigrants partly by the official way. There are reports two seed firms were offering 'Mammoth Russian' sunflower variety in their catalogues in the beginning of 1880, and that the U.S. consul in St. Petersburg sent cultivars to the USA in 1893 (Heiser 1976).

Russian immigrants to USA and Canada cultivated sunflower for roasting and eating seed (confectionary varieties). Other people planted it mainly to feed poultry and for silage (Putt 1997). Sunflower oil produced in a very limited scale.

One of the first sunflower oil breeding program was started in Canada in the middle of the 1930s. Mennonite cultivars were used as an initial material. Mammoth Russian was suitable for confectionary and silage production but not for oil crushing. Breeding purpose was to create early short-stem variety with small seeds rich with oil. Breeders used also other germplasm from Russia and USA. Sunflower acreage reaches 2000 ha in Canada by 1943 (Putt 1997).

The most dramatic event in sunflower evolution took place in a very modest place. Young agronomist started to teach agriculture the children of Kazaks in the small agricultural school of Ekaterinodar town (now it is the All-Russian Oil Crops Research Institute in Krasnodar) in the Soviet Union. His name was Vasiliy Stepanovich Pustovoit, and now he is famous among the sunflower breeders of the entire world. Pustovoit started to work with the sunflower from 1910. His main idea was to increase the oil content in the seeds. Many Russian scientists were sure that 33% of oil content is a biological limit for sunflower seed at that moment. But Pustovoit produced modern oil-type sunflower varieties with high oil content (up to 52% instead of 30-32% before) in the seeds. It allowed sunflower to become a world-wide important oil crop. For example Putt (1997) considered the introduction of cultivars to Canada from the USSR that had maturity levels comparable to Advent and Admiral, but much higher oil content as one of the two most important events in sunflower crop evolution on the American continent.

The first significant breeding programs to improve sunflower for North American agriculture were started in Saskatchewan by the Canada Department of Agriculture in 1937 and in Texas by the U.S. Department of Agriculture and Texas Agricultural Experiment Station in 1950 (Fick and Miller 1997). Smaller programs were also conducted in the 1950s by the Minnesota and California Agricultural Experiment Stations, and Northrup King and Company in California.

Breeding for resistance (or at least tolerance) to pests and diseases is always considered the most important aim in sunflower breeding. The list of most important diseases and pests depends on the region and time. Primarily in Russia, it included sunflower broomrape, rust and sunflower moth. Resistance to sunflower moth was achieved by screening all the material for a special black layer in the seed hull. First resistant local varieties appeared rather soon (Morozov 1947).

The next most important problem for sunflower in the European region became a broomrape. Broomrape (*Orobanche cumana* Wallr. syn. *O. cernua* Loefl.) is a parasitic plant, feeding on sunflower roots. It is interesting that sunflower being a plant from the American continent found this parasite in Europe. Broomrape is still absent in the sunflower's home land. The main method to control broomrape is the development of resistant varieties or hybrids. But broomrape constantly evolves and produces new races. During the over 100 years of co-evolution sunflower and broomrape in Russia race A was subsequently replaced by B and then E (now F, G and H also) (Gontcharov et al. 2004; Gontcharov 2009).

Breeding for resistance requires new genetic resources. Wild species, mainly *H. tuberosus*, were the main sources for introduction useful genes into susceptible sunflower (Pustovoit et al. 1976). The early former Soviet Union cultivars and *H. tuberosus* were also important sources of resistance for the broomrape complex of races in Romania (Vrânceanu et al. 1980). Early reports of broomrape resistance were from cultivars "Progress" and

"Novinka", which were developed using the "Group Immunity" breeding approach with germplasm derived from wild perennial *H. tuberosus* (Pustovoit and Gubin 1974). Immunity to broomrape in lines derived from *H. tuberosus* was also described (Pogorietsky and Geshle 1976; Seiler and Yan 2014).

Rust (*Puccinia helianthi* Schw.), being one of the most important sunflower pathogen previously was practically vanished after releasing of resistant cultivars in the 1960s. During a long time, rust was observed on foreign sunflower hybrids only (predominantly confectionary type or old samples from genetic collections). Now rust could be found on the leaves of modern oil-type sunflower hybrids. In 2012, rust was registered on all hybrids in the trial, though to a very small extent. It is possible that new races of rust appeared here (Gontcharov 2014).

All these achievements were made on the basis of the method of reserves. This widely used and highly successful method for improving sunflower cultivars was developed by Pustovoit during the 1920s. The method is a form of recurrent selection that includes progeny evaluation and subsequent cross pollination among progenies with superior characteristics (Fick and Miller 1997).

Successfully using the method of reserves, Soviet breeders developed several open-pollinated varieties with high oil content, yield and resistance to diseases and pests. Some varieties were so good that were introduced in other countries, including the USA and Canada. But the next evolutionary step for sunflower crop was nearby. It was made by Leclercq (1969; 1971) and Kinman (1970). Leclercq obtained cytoplasmic male sterility (CMS) by crossing cultivated sunflower with wild *Helianthus petiolaris*. Then Kinman discovered a fertility restorer gene for this cytoplasm. It allowed producing hybrid sunflower seeds on a commercial scale. After it F_1 sunflower hybrids with uniformity and high yield potential were developed and replaced OP-varieties on the fields throughout the world. Now only a few countries still use open-pollinated sunflower varieties for oil production.

Breeding technology changed, but the aims remained: as previously they were high yield, optimal vegetation period, oil content, adaptability and resistance to pests and diseases. Now, main efforts were directed to the inbred line development with desirable characteristics.

The development of a new cultivar, combining a short vegetation period with high productivity, is a rather challenging problem. Many researchers report correlations between yield and duration of the vegetation period for different crops, including sunflower (Putt 1943; Kovacik and Skaloud 1972; Stoenescu et al. 1985; Merrien 1992). At the same time, the possibility of a successful combination such traits in one hybrid is not completely denied (Pustovoit 1939). Breeding program at VNIIMK (Russia) aimed for development ultra-early inbred lines and hybrids was successful and a new sunflower hybrid was produced with vegetation period 75 days only and seed yield up to 3.5-4.0 t per ha.

Wild sunflower species are often used as a source of new resistance genes (Seiler 1992; Vear 2004; 2010; Vear et al. 2008a; Kaya et al. 2012).

Among the sunflower diseases downy mildew is one of the most important one for sunflower around the world. It caused by *Plasmopara halstedii* (Farl.) Berl. et de Toni (Novotelnova 1962; Goossen and Sackston 1968; Sackston 1992; Gulya et al. 1997; Jocić et al. 2012; Iwebor et al. 2016). There are few ways to control downy mildew, including agrotechnical, chemical and breeding methods. Seed chemical treatment and cultivation of sunflower hybrids, resistant to downy mildew are the most effective measures (Gulya et al. 1997; Vear et al. 1997). Breeding for resistance to downy mildew concentrates usually on the search and incorporation of major genes designated *Pl* into elite sunflower lines. Simple inheritance of the trait was discovered rather early (Vranceanu and Stoenescu 1970; Zimmer 1974), and different race-specific single dominant resistance genes are used worldwide by sunflower breeders (Vear et al. 2008a). This provokes downy mildew pathogen to produce new races (Ahmed et al. 2012). After race 100 and 300 appeared 710 and 730 and then 304, 307, 314, 334, 704 and 714, evolved (Carson 1981; Gulya 2007; Gulya et al. 1991b; Molinero-Ruiz et al. 1998; 2002; Tourvieille de Labrouhe et al. 2000). Now more than 20 Pl genes have been discovered and incorporated into commercial sunflower hybrids (Tourvieille de Labrouhe et al. 2008; Qi et al., 2016; Trojanova et al. 2017).

Division on races is based on the reaction of host (an internationally accepted set of differential lines) to the pathogen (Gulya et al. 1991a; Gulya 1995; Tourvieille de Labrouhe et al. 2000; 2012). Also, molecular methods are used in *P. halstedii* differentiation (Roeckel-Drevet et al. 2003; Gascuel et al. 2015) and in sunflower breeding for resistance (Mulpuri et al. 2009).

New races emergence forces sunflower growers to use chemical control by seed-applied fungicides. Usually seeds are treated by metalaxyl or mefenoxam (Melero-Vara et al. 1982; Albourie et al. 1998; Gulya 2002; Molinero-Ruiz et al. 2005). However, in 1995 metalaxyl tolerant isolates were discovered (Molinero-Ruiz et al. 2005; Spring et al. 2006) and a bit later – mefenoxam-tolerant isolates (Molinero-Ruiz et al. 2005).

Durable resistance could be achieved by combining race-specific (vertical) and non-race-specific (horizontal) resistance in one hybrid (Tourvieille de Labrouhe et al. 2004; Vear 2004; Vear et al. 2008b). To reach this, it is possible to use one parental line with the most effective for the specific location major gene and the second parental line – with a high horizontal resistance to downy mildew.

Among the relatively new diseases there is *Diaporthe* stem canker, which was seems to destroy all the sunflower acreages in the southern regions in Russia in 1997, but steadily decreased the pressure when susceptible varieties were steadily replaced with "stay green" hybrids. Meanwhile, susceptible lines show its symptoms every year in sunflower nurseries.

Fusarium wilt is one of the most important diseases for many crops. But for sunflower it was earlier considered being of less importance or minor disease (Gulya et al. 1997). During the last years, it became a serious problem for sunflower crop in Russia (Antonova et al. 2002). Special breeding program for resistance to Fusarium sp. started at VNIIMK in 2001 using the laboratory test developed in VNIIMK (Gontcharov et al. 2006). As it is perfectly known, success in resistance breeding leads to disequilibrium in the "host-parasite" system and stimulates parasite to evolve new races. Varying climatic environments favor different pathogens in different years. All this requires constant control of phytopathologic situation (Van der Plank 1968).

New market for sunflower oil was created at VNIIMK in the 1970s. K. Soldatov developed the first sunflower cultivar with the high level of oleic acid in the oil. Such sunflower oil was similar to olive oil by quality. Now high oleic sunflower hybrids are rather popular in some countries.

Sustainable market demand for confectionery sunflower seeds made VNIIMK initiate a special breeding program with the aim to develop modern confectionery open-pollinated varieties. Dr. S. Borodin with his colleagues released four OP varieties – SPK, Lakomka, Borodinskiy and Oreshek (Borodin 2003). Their seeds are close to the oil-type one by structure but larger in size and 1000-seed weight, has bigger husk content and less oil content (450-490 g/kg). Husk is black or black with grey stripes in color. This type of seeds has a special Russian name "mezheumok" and means intermediate. People in Russia and Ukraine prefer such types of sunflower seeds for the direct consumption. Such seeds also could be easily dehulled by the machinery for confectionery use. Now these three OP varieties covered about 500 thousand hectares in Russia. Commercial success of confectionery OP varieties encouraged VNIIMK to begin a confectionery hybrid breeding program. This program started in 1999. Russian market demands for sunflower with 1000-seed weight 80 g or more with oil content on the level 450-490 g/kg and seeds should be easily dehulled (Borodin 2003). As a new initial breeding material we used non-oil samples of sunflower from Iran and Syria, Russian modern confectionery OP varieties and high-oil inbred lines of our breeding with relatively big seed size. As a result of crossing this material and self-pollination of obtained hybrids, we developed several inbred lines for confectionery hybrid breeding. Lines were crossed with CMS-lines to test their ability to restore pollen fertility. So we found some restorer lines and some maintainer lines. Several of such lines were converted to CMS-lines by back-crossing. The aim was to develop and evaluate new confectionery inbred lines and hybrids. New breeding program with the aim to develop sunflower hybrids suitable for double-using – as a confectionery and oil-type hybrid – started at VNIIMK in 1999. No such commercial hybrids were available at the moment in Russia. Such hybrids should perform high seed yield, rather high oil content and high 1000-seeds weight. Results show that two best hybrids (VD-354 A

× K-4 and VK-905 A × K-4) have significantly higher seed yield in comparison with the check. The seed yield level was rather high in the trial with check variety Oreshek yield 3.23 t/ha.

Oil content in the absolutely dry seeds was 454 g/kg for the check variety and varied from 431 to 480 g/kg in the seeds of the studied hybrid combinations. As a result tested hybrids could be used in two different ways (for oil production and for confectionery use) along with released confectionery OP varieties (Oreshek, SPK and Lakomka). Oil-type sunflower hybrids usually have a higher oil content (480-520 g/kg), but in this case industry had a problem with dehulling procedure. Significantly less oil content is typical for the confectionery sunflower produced outside Russia, but such material had no commercial success here.

The 1000-seeds weight of all tested hybrid combinations was higher 80 g, thought superiority of check variety was obvious. Comparison of 1000-seeds weight of all tested combinations showed big variation for this trait. 1000-seed weight varied from 79.1 g (VD-354 × K-3) to 109.9 g (VK-905 × K-3).

It was proved that to develop sunflower hybrids suitable for double-using – as a confectionery and oil-type hybrid – is possible. Such hybrids should perform high seed yield, rather high oil content and high 1000-seeds weight. Three-year trial allows us to define the most prominent hybrid combination Katyusha (VK-905 A × VK-944) (Gontcharov and Zaharova 2008).

In 2019, sunflower was planted in Russia on the area of more 7 million ha, with about 1 million ha – for confectionary usage.

The modern period in sunflower breeding is characterized by the application of up-to-date techniques – molecular tools, marker-assisted breeding et cetera.

REFERENCES

Ahmed, S., Tourvieille de Labrouhe, D. & Delmotte, D. (2012). Emerging virulence arising from hybridisation facilitated by multiple introductions of the sunflower downy mildew pathogen *Plasmopara halstedii*. *Fungal Genetics and Biology*, *49*, 847–55.

Albourie, J. M., Tourvieille, J. & Tourvieille de Labrouhe, D. (1998). Resistance to metalaxyl in isolates of the sunflower pathogen *Plasmopara halstedii*. *European Journal of Plant Pathology*, *104*, 235–42.

Borodin, S. G. (2003). Sunflower OP varieties breeding for specific use. p. 15-25. In: *Proc. Internat. Conf. devoted to 90 years of VNIIMK*, Krasnodar. (in Russian).

Burke, J. M., Tang, S., Knapp, S. J. & Rieseberg, L. H. (2002). Genetic analysis of sunflower domestication. *Genetics*., Vol. *161*, p. 1257-1267.

Carson, M. L. (1981). New race of *Plasmopara halstedii* virulent on resistant sunflowers in South Dakota. *Plant Disease*, *65*, 842–3.

Fick, G. N. & Miller, J. F. (1997). Sunflower breeding//Sunflower technology and production. *Agronomy*, *35*. /Schneiter A.A., editor. USA, Madison, p. 809-824.

Gascuel, Q., Martinez, I., Boniface, M. C., Vear, F., Pichon, M. & Godiard, L. (2015). The sunflower downy mildew pathogen *Plasmopara halstedii*. *Molecular Plant Pathology*, *16*, 109–22.

Gontcharov, S. V., Antonova, T. S. & Araslanova, N. M. (2004). Sunflower breeding for resistance to the new broomrape race. *Helia*., *27* (40), 193-198.

Gontcharov, S. V., Antonova, T. S. & Saukova, S. L. (2006). Sunflower breeding for resistance to *Fusarium*. *Helia*, *29*(45), 49-54.

Gontcharov, S. V. & Zaharova, M. V. (2008). Vegetation period and hybrid sunflower productivity in breeding for earliness., p. 531-533. In: *Proc. 17th International Sunflower Conference*. Cordoba. Spain.

Gontcharov, S. V. (2014). Dynamics of hybrid sunflower disease resistance. *Helia*, *37* (60), 99-104.

Gontcharov, S. V. (2009). Sunflower breeding for resistance to the new broomrape race in the Krasnodar region of Russia. *Helia*, *32*(51), 75–80.

Goossen, P. G. & Sackston, W. E. (1968). Transmission and biology of sunflower downy mildew. *Canadian Journal of Botany*, *46*, 5–10.

Gulya, T. J. (1995). Proposal of a revised system of classifying races of sunflower downy mildew. In: *Proceedings of 17th Sunflower research Workshop*. Fargo, ND, USA: National Sunflower Association, 76–78. [http://www.sunflowernsa.com/uploads/research/686/1995_gulya_proposalrevisedsystem.pdf]. Accessed 29 February 2016.

Gulya, T. J. (2002). Efficacy of single and two-way fungicide seed treatments for the control of metalaxyl-resistant strains of *Plasmopara halstedii* (sunflower downy mildew). In: *Proceedings of the BCPC Conference Pests and Diseases*. Brighton, UK: BCPC, 575–80.

Gulya, T. J. (2007). Distribution of *Plasmopara halstedii* races from sunflower around the world. In: Lebeda A, Spencer-Phillips PTN, eds. *Advances in Downy Mildew Research*, Vol *3*. Proceedings of the 2nd International Downy Mildews Symposium at Palacky University in Olomouc and JOLA, Kostelecna Hane, Czech Republic. Dordrecht, Netherlands: Springer, 121–34.

Gulya, T. J., Miler, J. F., Viranyi, F. & Sackston, W. E. (1991a). Proposed internationally standardized method for race identification of *Plasmopara halstedii*. *Helia*, *14*, 11–20.

Gulya, T. J., Sackston, W. E., Viranyi, F., Masirevic, S. & Rashid, K. Y. (1991b). New races of the sunflower downy mildew pathogen (*Plasmopara halstedii*) in Europe and North and South America. *Journal of Phytopathology*, *132*, 303–11.

Gulya, T. J., Rashid, K. Y. & Masirevic, S. M. (1997). Sunflower diseases. In: Schneiter A. A., ed. Sunflower Technology and Production, No. 35. Agronomy. Madison, WI, USA: *American Society of Agronomy*, 263–379.

Gundaev, A. I. (1971). Basic principles of sunflower selections, p. 417-465. *In Genetic principles of plant selection.* (In Russian.) Nauka, Moscow.

Heiser, C. B. (1951). The sunflower among the North American Indians. *Proc. Am. Phil. Soc.*, *95*, 432-448.

Heiser, C. B. (1955). Origin and development of the cultivated sunflower. *Am. Biol. Teach.*, 17, 161-167.

Heiser, C. B. (1976). *The sunflower*. Univ. Oklahoma Press, p. 1-198

Heiser, C. B., Smith, D. M., Clevenger, S. B. & Martin, W. C. (1969). The North American sunflowers (*Helianthus*). Memoirs Torrey *Bot. Club*, *22*(2), 1-218.

Heiser, C. B. (1978). Taxonomy of *Helianthus* and origin of domesticated sunflower//Sunflower science and technology/Carter J. F., editor. *Agronomy*, *19*. USA, Madison., p. 31-53.

Iwebor, M., Antonova, T. S. & Saukova, S. (2016). Changes in the racial structure of *Plasmopara halstedii* (Farl.) Berl. et de Toni population in the South of the Russian Federation. *Helia*, *39* (64), 113-121.

Jan, C. C. & Seiler, G. J. (2007). Sunflower. In: *RJ Singh (ed) Genetics Resources, Chromosome Engineering, and Crop Improvement*, vol *4*, Oilseed Crops. CRC Press, NY, USA, pp. 103–165.

Jocić, S., Miladinović, D., Imerovski, I., Dimitrievic, A., Cvejić, S., Nagi, N. & Kondic-Spika, A. (2012). Towards sustainable downy mildew resistant in sunflower. *Helia*, *35* (56), 61-72.

Kaya, Y., Jocic, S. & Miladinovic, D. (2012). Sunflower. In: *S.K. Gupta (ed.), Technological Innovations in Major World Oil Crops*, V. *1*, 85-129.

Kinman, M. L. (1970). New developments in the USDA and state experiment station sunflower breeding programs p 181-183 *Proc 4th Int. Sunflower Conf*, Memphis, TN 23-25 June. Int. Sunflower Assoc, Pans, France

Kovacik, A. & Skaloud, V. (1972). The proportion of the variability component caused by the environment and the correlations of economically important properties and characters of the sunflower (*Helianthus annuus* L.)., p. 249 – 261. In: *Science Agricultural Bohemoslov*, № 4.

Leclercq, P. (1969). Une sterilite male cytoplasmique chez le tournesol [Cytoplasmic male sterility in sunflowers]. *Ann. Amelior. Plantes*, *19*, 99-106.

Leclercq, P. (1971). La sterilite male cytoplasmique du tournesol I Premieres etudes sur la restauratm de la fertihte [Cytoplasmic male sterility of the sunflower I First studies on the restoration of fertility]. *Ann. Amelior. Plant*, *21*, 45-54.

Lentz, D. L., Pohl, M. E. D., Pope, K. O. & Wyatt, A. R. (2001). Prehistoric sunflower (*Helianthus annuus* L.) domestication in Mexico//*Economic Botany*., V. *55*(3), P. 370-376.

Merrien, A. (1992). *Some aspect of sunflower crop physiology*., p. 481 – 498. In: Proc. 13th Conf., Pisa, Italy, *Int. Sunf. Assoc*.. V. *1*.

Melero-Vara, J. M., Garcia-Baudin, C., Lopez-Herrera, C. J. & Jimenez-Diaz, R. M. (1982). Control of sunflower downy mildew with metalaxyl. *Plant Disease*, *66*, 132–5.

Molinero-Ruiz, L., Melero-Vara, J. M. & Gulya, T. J. (1998). Pathogenic characterization of *Plasmopara halstedii* isolates from Spain. In: Gulya TJ, Vear F, eds. *Proceedings of the International Sunflower Association Symposium III, Sunflower Downy Mildew*. Fargo, ND, USA: International Sunflower Association, 26–9.

Molinero-Ruiz, M. L., Dominguez, J. & Melero-Vara, J. M. (2002). Races of isolates of *Plasmopara halstedii* from Spain and studies on their virulence. *Plant Disease*, *86*, 736–40.

Molinero-Ruiz, M. L., Dominguez, J., Gulya, T. J. & Melero-Vara, J. M. (2005). Reaction of field populations of sunflower downy mildew (*Plasmopara halstedii*) to metalaxyl and mefenoxam. *Helia*, *28*, 66–74.

Morozov, V. K. (1947). *Sunflower breeding in the USSR*., 271 p. Moscow. (in Russian).

Mulpuri, S., Liu, Z., Feng, J., Gulya, T. J. & Jan, C. C. (2009). Inheritance and molecular mapping of a downy mildew resistance gene *Pl*13 in cultivated sunflower (*Helianthus annuus* L.). *Theoretical and applied Genetics*, *119*, 795–803.

Pogorietsky, P. K. & Geshle, E. E. (1976). Sunflower immunity to broomrape and rust. In: *Proc. 7th Intl. Sunfl. Conf*., Krasnodar, Russia. Intl. Sunfl. Assoc., Paris, France, 27 June–3 July, pp. 238–243.

Pustovoit, V. S. (1939). *Sunflower breeding for increased oil content, breeding methods, results and prospects*. (PhD thesis). Krasnodar, Russia. (in Russian).

Pustovoit, V. S. (1964). Conclusions of work on the selection and seed production of sunflowers. (In Russian). *Agrobiology*, *5*, 662-697.

Pustovoit, G. V. & Gubin, I. A. (1974). Results and prospects in sunflower breeding for group immunity by using the interspecific hybridization method. In: *Proc. 6th Intl. Sunfl. Conf.*, Bucharest, Romania. Intl. Sunfl. Assoc., Paris, France, 22–24 July, pp. 373–381.

Pustovoit, G. V., Ilatovsky, V. P. & Slyusar, E. L. (1976). Results and prospects of sunflower breeding for group immunity by interspecific hybridization. In: *Proc. 7th Intl. Sunfl. Conf.*, Krasnodar, Russia. Intl. Sunfl. Assoc., Paris, France, 27 June–3 July, pp. 193–204.

Putt, E. D. (1943). Association of seed yield and oil content with other characters in the sunflower. *Science Agricultural.*, *23*, 377 – 382.

Putt, E. D. (1997). Early history of sunflower. p. 1-19. In: A.A. Schneiter (ed.), Sunflower Production and Technology. *Agronomy Monograph*, *35*. ASA, CSSA and SSSA, Madison, WI, USA.

Qi, L. L., Foley, M. E., Cai, X. W. & Gulya, T. J. (2016). Genetics and mapping of a novel downy mildew resistance gene, *Pl*18, introduced from wild *Helianthus argophyllus* into cultivated sunflower (*Helianthus annuus* L.). *Theoretical and Applied Genetics*, *129*, 741–52.

Roeckel-Drevet, P., Tourvieille, J., Gulya, T. J., Charmet, G., Nicolas, P. & Tourvieille de Labrouhe, D. (2003). Molecular variability of sunflower downy mildew, *Plasmopara halstedii*, from different continents. *Canadian Journal of Microbiology*, *49*, 492–502.

Sackston, W. E. (1992). On a treadmill: breeding sunflowers for resistance to disease. *Annual Review of Phytopathology*, *30*, 529–51.

Schilling, E. E. & Heiser, C. B. (1981). Infrageneric classification of *Helianthus*. *Taxon*, *30*, 393–403.

Seiler, G. J. (1992). Utilization of wild sunflower species for the improvement of cultivated sunflower. *Field Crops Research*, *30*, 195–230.

Seiler, G. J. & Jan, C. C. (2014). Wild Sunflower Species as a Genetic Resource for Resistance to Sunflower Broomrape (*Orobanche cumana* Wallr.) *Helia.*, *37*(61), 129–139.

Semelczi-Kovacs, A. (1975). Acclimatization and dissemination of the sunflower in Europe. (In German). *Acta Ethnogr. Acad. Sci. Hung.*, *24*, 47-88.

Severin, G. (1939). Sunflower seed and sunflower oil. p. 77-89. In Studies of principal agricultural products of the world market. No. 4. Oils and fats: Production and international trade. Part I. *Int. Inst. Agric*, Rome, Italy.

Spring, O., Zipper, R. & Heller-Dohmen, M. (2006). First report of metalaxy lresistant isolates of *Plasmopara halstedii* on cultivated sunflower in Germany. *Journal of Plant Diseases and Protection*, *113*, 224.

Stebbins, J. C., Winchell, C. J. & Constable, J. V. H. (2013). *Helianthus winteri* (Asteraceae), a new perennial species from the southern Sierra Nevada foothills, California. *Aliso*, *31*, 19–24.

Stoenescu, F., Pârvu, N., Iuoraş, M., Terbea, M. & Voinescu, G. (1985). Particularităti ale ameliorării florii-soarelui pentru optimizarea periodadei de vegetatie [Particularities of sunflower enhancement for the optimization of the vegetation period]., p. 219 – 240 In: *Probleme de genetică teoretică şi aplicată.*, Vol. *XVII.*

Tourvieille de Labrouhe, D., Gulya T. J., Masirevic, S., Penaud, A., Rashid, K. & Viranyi, F. (2000). New nomenclature of races of *Plasmopara halstedii* (sunflower downy mildew). In: *Proceedings of the 15th International Sunflower Conference*. Toulouse, France: International Sunflower Association, I 61–6.

Tourvieille de Labrouhe, D., Walser, P., Mestries, E., Penaud, A., Tardin, M. C. & Pauchet, I. (2004). Sunflower downy mildew resistance gene pyramiding, alternation and mixture: first results comparing the effects of different varietal structures on changes in the pathogen//*Proc. 16th Int. Sunflower Conf.*, Fargo, USA., 255.

Tourvieille de Labrouhe, D., Serre, F., Walser, P., Roche, S. & Vear, F. (2008). Quantitative resistance to downy mildew (*Plasmopara halstedii*) in sunflower (*Helianthus annuus*). *Euphytica*, *164*, 433–44.

Trojanova, Z.., Sedlarova, M., Gulya, T. J. & Lebeda, A. (2017). Methodology of virulence screening and race characterization of *Plasmopara halstedii*, and resistance evaluation in sunflower – a review. *Plant Pathology.*, *66* (2), 171-185.

Van Der Plank, J. E. (1968). *Disease Resistance in Plants*, Academic Press, New York and London, pp. 248.

Vear, F. (2004). Breeding for durable resistance to the main diseases of sunflower//*Proc. 17th Int. Sunflower Conf.*, USA, Fargo., 125-130.

Vear, F. (2010). Classic genetics and breeding. In: Hu J, Seiler G, Kole C, eds. *Genetics, Genomics and Breeding of Sunflower*. Enfield, NH, USA: Science Publishers, 51–78.

Vear, F., Gentzbittel, L., Philippon, J., et al. (1997). The genetics of resistance to five races of downy mildew (*Plasmopara halstedii*) in sunflower (*Helianthus annuus* L.). *Theoretical and Applied Genetics*, *95*, 584–9.

Vear, F., Serieys, H., Petit, A., et al. (2008a). Origins of major genes for downy mildew resistance in sunflower. In: *Proceedings of 17th International Sunflower Conference.* Cordoba, Spain: Consejeria de Agricultura y Pesca, 125–30.

Vear, F., Serre, F., Jouan-Dufournel, I., et al. (2008b). Inheritance of quantitative resistance to downy mildew (*Plasmopara halstedii*) in sunflower (*Helianthus annuus* L.). *Euphytica*, *164*, 561–70.

Vranceanu, V. & Stoenescu, F. (1970). Immunity to sunflower downy mildew due to a single dominant gene. *Probleme Agriculture*, *22*, 34–40.

Wills, D. M. & Burke, J. M. (2007). Quantitative trait locus analysis of the early domestication of sunflower. *Genetics*, *176*, 2589-2599.

Zimmer, D. E. (1974). Physiological specialization between races of *Plasmopara halstedii* in America and Europe. *Phytopathology*, *64*, 1465–7.

Zukovsky, P. M. (1950). *Cultivated plants and their wild relatives.* Common. Agric. Bur., Franham Royal, England.

In: Sunflowers
Editor: Érico de Sá Petit Lobão
ISBN: 978-1-53617-195-2

Chapter 2

SUNFLOWER CROP UNDER ORGANIC FERTILIZATION IN BRAZIL

Érico de Sá Petit Lobão*, Albericio Pereira Andrade, Pedro Fernandes Dantas, Javob Silva Souto, José Marques Pereira and Dan Érico Lobão
Pau Brasil Foundation, Itabuna, Brazil

ABSTRACT

In Brazil, research on sunflower (*Hellianthus annnus* L.) was initiated by the Campinas Agronomic Institute (IAC) of the State of São Paulo, whose first records date from 1932. However, the interest and increase of sunflower cultivation in Brazil occurred mainly due to research and technological development in the sector in the 90s. In addition, the numerous forms of use, the emergence of industries and the need for farmers to diversify their crops, whether in the form of oil for human consumption or for the production of biodiesel, in animal feed through bird seed, bran, pie or silage for ruminants and monogastric, have contributed to increase research and production, as well as for the establishment of its production chain in the different regions of the country. Besides the great challenges for sunflower production and use, nitrogen deficiency has been

* Corresponding Author's Email: ericolobao@hotmail.com.

identified as the most frequent nutritional disorder of the plant causing growth problems and plant stem breakage. Organic fertilization brings physical, chemical and biological benefits. When it comes to organic fertilizers, animal manures are the most important, whether for their composition, relative availability or application benefits.

Keywords: *Helianthus annuus*, cattle manure, sunflower in Brazil

SUNFLOWER IN BRAZIL

In Brazil, research on sunflower (*Hellianthus annnus*) was initiated by the Campinas Agronomic Institute (IAC) of the State of São Paulo, whose first records date from 1932 (Ungaro 2000). In Rio Grande do Sul, research began in the 1950s. Due to the demand for energy sources, in 1980, experimental work was restarted, supported by the existence of hybrids and research information carried out especially in the states of Paraná and São Paulo.

The works were basically developed in the areas of crop management and genetic improvement (Leite et al. 2005). However, the interest and increase of sunflower cultivation in Brazil occurred mainly due to research and technological development in the sector in the 90s. In addition, the numerous forms of use, the emergence of industries and the need for farmers to diversify their crops, whether in the form of oil for human consumption or for the production of biodiesel, in animal feed through bird seed, bran, pie or silage for ruminants and monogastric, have contributed to increase research and production, as well as for the establishment of its production chain in the different regions of the country (Oliveira 2001).

The sunflower is one of the fastest growing oilseeds in recent years, both in cultivation area and in production, and is currently classified as the second largest source of raw material for the edible oil industry in the world (Souza et al., 2005). Sunflower is currently among the five largest edible vegetable oil-producing oilseeds in the world and the fifth largest in cultivated area in the world, covering an area of approximately 20 million hectares. During the 2010/11 harvest, its production was 7% of world oilseed production, second only to soybean, canola, cotton and peanuts, producing the equivalent of 12%, 52%, 72% and 87% respectively of these crops (CONAB 2012).

Brazil is the world's second largest producer of oilseeds, but the largest sunflower producers are Ukraine, Russia and Argentina (USDA 2011). The area of sunflower harvested worldwide during the 2010/11 harvest was approximately 2.24 million hectares (USDA 2011). In Brazil, the area harvested during the 2007 harvest was 73.2 thousand hectares, with a production of 104.9 thousand tons. The states of Goiás, Mato Grosso, Mato Grosso do Sul and Rio Grande do Sul are the largest producers, accounting for 93% of Brazilian production. The sunflower planted area in Brazil, from 1998 to 2007, grew from 12,400 ha to 73,200 ha, which represents an increase of about 590%, i.e., almost 100% per year (IBGE, 2008). In the Second Planting Intention Survey," conducted by Conab, in November 2012, it was estimated that the area of sunflower cultivation for the 2012/13 crop should be maintained. The cultivated area was around 74.5 thousand hectares, with 47.1 thousand hectares cultivated in Mato Grosso, corresponding to 63%. Then comes the state of Goiás with 13.9 thousand hectares, or 19% of the national area sown with sunflower. Estimates of national sunflower production for the 2012/13 crop were around 93.6 thousand tons, 20% lower than the previous crop (CONAB 2012).

In research on sunflower cultivation it is very important to identify genotypes that have favorable characteristics of grain yield, disease tolerance, cycle, oil content and adaptation to mechanized harvesting (Trezzi et al. 1997). The evaluation and selection of sunflower genotypes is being done through an official testing network. This network has the participation of public and private institutions, being coordinated by Embrapa Soy. The trials have been conducted in several locations of Rio Grande do Sul, Santa Catarina, Paraná, Sao Paulo, Goiás, Maranhão, Mato Grosso, Mato Grosso do Sul, Minas Gerais, Bahia, Tocantins and Federal District (Porto 2006).

The basically three challenges facing sunflower in Brazil are: I) to offer producers an alternative crop that produces two harvests per year; II) to offer one more oilseed raw material to the other grain processing industries, reducing their idleness; and III) finally offer the market an edible oil of high nutritional value (Pelegrini, 1985). Added to these challenges is the current alternative of energy production, since sunflower oil can be used as a raw material for biodiesel production (Leite et al., 2005), as well as the use of its agro-industrial waste, such as animal feed, as the increase in sunflower production in the country has paved the way for the use of sunflower biomass, not just for oil production and other usual uses.

As a result of the national biodiesel program, sunflower cultivation for this purpose has been growing every year in Brazil. As of 2008, the program includes 3% biodiesel in diesel oil, representing approximately one billion liters of biodiesel per year, which has been reaching the production of oilseeds with desirable characteristics for this purpose (Lima 2011).

Technology is available to guarantee or develop sunflower production for different Brazilian regions, under favorable conditions, in terms of physical yield per hectare. Its insertion in the production chain is also assured, considering that it uses the same structure available for soybeans (Monteiro 2007).

The sunflower cultivation in Brazil is relatively recent, therefore, due to the interaction between the model and the environment, present in plant species, the continuous evaluation of sunflower genotypes is necessary (Porto 2006). Based on the results obtained by the Sunflower Genotype Evaluation Testing Network, coordinated by Embrapa Soy, observe if there is great potential for its production in the states of São Paulo, Rio Grande do Sul, Santa Catarina, Federal District, Mato Grosso, Mato Grosso do Sul, Goiás, Minas Gerais, Tocantins, Bahia, Maranhão and Piauí. As regions that are quite distinct in relation to climate, soil and land structure, they are characterized as grain producing areas because they have the necessary vulnerability to sunflower production (Oliveira and Vieira 2004).

For the medium and large rural producer the sunflower crop fulfills the option of crop rotation and succession with advantages over other plants. For the small producer, the grains serve for poultry feed and human consumption. In addition, the existence of an oil extraction micro-machine, accessible to cooperatives, producer associations and even medium-sized farmers, allows the extraction of cold oil, which serves both medicinal and domestic purposes, on the property or local market (Monteiro 2007).

SUNFLOWER CHARACTERIZATION

The sunflower is a crop of short cycle, high quality and high oil yield, characteristics that make it a good option for Brazilian producers (Gontijo-Neto et al. 2009). It presents root system with pivoting main root, its leaves are alternated and petiolate. Inflorescence is a chapter in which grains, called achenes, develop. The stem and the chapter are the components with the largest participation in sunflower mass production (Tomich et al. 2003).

In addition, sunflower is the oilseed with greater tolerance to drought, cold and heat, when compared with most species cultivated in Brazil. These authors also report that the sunflower adapts well to changing temperature conditions, considering the range between 18°C and 24°C as the best for crop development. During the early stages of its cycle (0 to 40 days), the plant has tolerance to low temperatures and drought, and in the following phases, excessive cold and lack of water cause changes in the plants, causing loss of production. It requires deep, well-drained, fertile soils, preferably clayey, with good nitrogen, phosphorus and potassium supplies for high yields. However, the crop also has the ability to grow in less fertile soils with poor physical characteristics, provided that the necessary minimum corrections are made (Leite et al. 2005).

The duration of the growing season can vary from 90 to 130 days, depending on the materials used for cultivation, from the earliest to the latest, the date of sowing and the environmental conditions characteristic of each region and year. However, there is no consensus on the adaptability of cultivars, nor fixed sowing periods for different Brazilian regions. Thus, choosing the planting period becomes crucial for the success of the crop. In Brazil, the existing recommendations for the different regions indicate two possible seasons for sowing, summer and autumn (Costa et al., 2000 Porto 2006).

The sunflower can be grown in harvest (August/September sowing) in Rio Grande do Sul and Paraná, and in safrinha (February/March sowing) in São Paulo, Mato Grosso do Sul, Mato Grosso, Goiás, Federal District, Bahia and Maranhão (Porto 2006). According to Úngaro (2000), the sunflower must be harvested when the water content of the grain is between 14% and

16%, since with higher moisture content they can stain and acquire odors that pass to the oil. In this case, it is advisable to dry in yards or in a dryer. In early cultivars this will occur around 100 days and in late cultivars around 120 days after plant emergence, depending on the climatic conditions of the region.

Sunflower Organic Fertilization

According to Favara et al. (2009), in general, the requirement of sunflower for soil fertility is similar to that of soybean and corn, but does not tolerate soil acidity and compaction, which may limit its development, intensifying the nutritional problems associated with deficit and reducing the productive potential of the crop. According to the authors, sunflower promotes improvement in soil fertility by nutrient cycling and still observing low nutrient exports. The optimum temperature for sunflower crop development is around 27°C and the soil must have a pH above 5.2 to avoid toxicity.

According to Úngaro (2000), nitrogen deficiency has been identified as the most frequent nutritional disorder of the plant, and phosphorus and potassium poor soils cause growth problems and plant stem breakage.

Organic fertilization brings physical, chemical and biological benefits. When it comes to organic fertilizers, animal manures are the most important, whether for their composition, relative availability or application benefits (Souto et al. 2005). The benefits of animal manure use outweigh improvements in soil physical properties and nutrient supply, increased organic matter content, improved water infiltration as well as increased cation exchange capacity (Hoffman et al. 2001).

According to Oliveira et al. (2009), high manure contents may provide nutritional imbalance in the soil and, consequently, reduced development and final yield of sunflower crop. For Rossi (1998), cattle manure significantly increases yields in years with adequate rainfall and soil moisture in sunflower crop. In addition, the adoption of organic manure with cattle manure, among others, becomes a viable alternative due to its ease of obtaining and the relatively low cost (Nobre et al. 2010).

Among the nutrients that make up the manure, nitrogen stands out for playing an important role in the metabolism and nutrition of sunflower crop, and its deficiency causes nutritional disorder, and this nutrient is the one that most limits its production, while Excess causes a decrease in the percentage of oil, and high doses may increase the incidence of pests and diseases, affecting grain production (Biscaro et al. 2008).

Regarding sunflower fertilization, it has been observed that the crop accumulates large amounts of nutrients, mainly nitrogen, phosphorus and potassium. Its deep root system provides further exploration and assists in better utilization of the natural fertility of the soils and fertilizers of previous crops, absorbing nutrients from deeper layers (Santos and Grangeiro 2013). However, Santos et al. (2010) warns that boron (B) is a nutrient found in low concentrations in the plant, being essential for plant development, and its deficiency often causes nutritional problems in sunflower crop.

Nitrogen fertilization recommendations in sunflower range from 40 to 80 kg ha-1 N. Because this nutrient is extracted by the crop in large quantities and has no direct residual effect on the soil, the expected yield is a function of N dosages. used (Lobo et al., 2011). Experimental results indicate that the maximum sunflower yield was reached with 80 to 90 kg ha-1 N, however, applying 40 to 50 kg ha-1 N yields 90% of the maximum relative yield, corresponding to the fertilization levels. more economically efficient (Smiderle et al. 2009).

In my theses, studing, the *Hellianthus annus* cv. Helio 250, I found that fertilization with upland cattle manure has no effect on dry matter and organic matter yield of sunflower cake provided an increase in crude protein and mineral matter contents and a reduction in the fiber content of this co-product. 2. The most promising level of inclusion of manure was 22.5 t ha-1, however, for recommendation of organic fertilization, doses of cattle manure between 13 and 23 t ha-1 provide better results because they have higher levels. of crude protein and lower fibrosity of sunflower cake. 3. In addition, the adoption of organic fertilization of sunflower with cattle manure is a viable alternative for improving sunflower cake quality, mainly due to the ease of obtaining manure and the relatively low cost (Lobão et al. 2015). The addition of manure in the cultivation of sunflower cv. 250 helium increases the productivity of seed, without observing effects of doses, provided there are no limitations of water in the soil. The cattle manure fertilization has no effect on the oil content of the seeds obtained nor pie, but led to an increase in the productivity of oil and consequently the pie. The maximization of grain and pie production can be achieved with organic manure of bovine in 37.5 t ha^{-1}, what means a fertilization with 352.5 kg N ha-1 (Lobão et al. 2017).

Acknowledgments

To the staff of the Animal Nutrition Laboratory at the Center for Agricultural Sciences, Charles, Duelo, Mr Costa, José Sales and Roberto to the cowboys Lendro and Cristiano (Pio). To my colleagues Adelilian Baracho, Daniely Sales, Meyre Cassuce, Paula Frassinetti, and graduate students Ana Paula Brito, Elton Silva, Valdiléia Avelar and Luciana Firmino. To the teachers, Elizanilda Ramalho do Rego and Walter Esfrain Pereira. To the company Helianthus do Brasil for the donation of the seeds of the cultivar Helium 250.

References

Biscaro, G. A., Machado, J. R., Tosta, M. S., Mendonça, V., Soratto, R. P. and Carvalho, L. A. (2008). Covered nitrogen fertilization in irrigated sunflower under the conditions of Cassilândia-MS. *Ciência e Agrotecnologia*, Lavras, 32 (5): 1366-1373.

Companhia Nacional de Abastecimento - CONAB. (2012). *Brazilian crop monitoring: grain, according to survey*, November 2012/Companhia Nacional de Abastecimento. Brasília: Conab.

Costa, V. C. A., Silva, F. N. and Ribeiro, M. C. C. (2000). Effect of sowing dates on sunflower germination and development (*Helianthus annuus* L.). *Revista Científica Rural*, 5: 154-158.

Gontijo-Neto, M. M., Leite, C. E. P., Uba, M. A. et al. (2009). Evaluation of sunflower and perennial tropical forages in intercropping. *Embrapa Corn and Sorghum (Boletim de Pesquisa e Desenvolvimento 19).*

Hoffmann, I., Gerling, D., Kyiogwom, U. B. and Mané-Bielfeldt, A. (2001). Farmers management strategies to maintain soil fertility in a remote area in northwest Nigeria. *Agriculture, Ecosystems & Environment*, 86: 263-275.

Leite, R. M. V. B. C., Brighenti, A. M. and Castro, C. (2005). *Sunflower in Brazil*. Londrina: Embrapa Soy.

Lima, H. L. (2011). *Nutritional parameters in steers supplemented with sunflower pie grazing Brachiaria brizantha cv. Marandu. 2011. 89p. Dissertation (Master in Zootecnia)*. Faculdade de Ciências Agrárias, UFGD, Dourados, MS.

Lobão, E. S. P., Andrade, A. P., Fernandes, P. D., Medeiros, E. P., Santos, E. M., Souto, J. S. and and Lobão, D. E. (2017). Morphological Characterization of Sunflower under Organic Fertilization and Seed Oil Content and Yield Pie. *Helia*. 40 (66): 29-45. Published Online: 2017-02-07 | doi: https://doi.org/10.1515/helia-2016-0001.

Lobão, E. S. P., Andrade, A. P., Fernandes, P. D., Silva, D. S., Gonzaga-Neto, S., Medeiros, A. N. and Lobão, D. E. (2015). Bromatological attributes of sunflower cake under organic fertilization. *Agrotrópica*, Ilhéus. 27 (3): 281-288.

Lobo, T. F., Grassi-Filho, H. and Brito, I. C. A. (2011). Effect of nitrogen on sunflower nutrition. *Bioscience Journal*, Uberlândia. 27 (3): 380-391, 2011.

Monteiro, J. M. G. (2007). *Planting oilseeds by family farmers from the semiarid Northeast for biodiesel production as a climate change mitigation and adaptation strategy*. 2007. 315p. Theses (Doctor in Energy Manager Sciences), Universidade Federal do Rio de Janeiro, UFRJ, Rio de Janeiro, RJ.

Nobre, R. G., Gheyi, H. R., Soares, F. A. L., Andrade, L. O., Nascimento, E. C. S. (2010). Sunflower production under different blades with domestic effluents and organic fertilization. *Revista Brasileira de Engenharia Agrícola e Ambiental*, 14 (7): 747-754.

Oliveira F. A., Oliveira-Filho, A. F., Medeiros, J. F. et al. (2009). Initial development of castor bean under different sources and doses of organic matter. *Revista Caatinga*, 22 (1): 206-211.

Oliveira, M. F. (2001). *Technologies for sunflower crop development in Brazil. In: Search results for Embrapa Soy - 2000: sunflower and wheat.* Hoffmann-Campo, C. B. e Saraiva, O. F. (Org.). Londrina: Embrapa Soy (Documentos Embrapa 165).

Oliveira, M. F. and Vieira, O. V. (2004). *Sunflower oil extraction using mini press*. Embrapa, Londrina (Documentos 237).

Porto, W. S. (2006). *Selection criteria of sunflower (Helianthus annuus L.) Genotypes evaluated in different environments. 100p. Dissertation (Master in Genetic and Genetic Improvement).* Universidade Estadual do Maringá. UEM, Maringá, PR.

Rossi, R. O. (1998). *Suwflower*. Curitiba: Editora Tecnoagro LTDA. 333p.

Santos, J. F. and Grangeiro, J. I. T. (2013). Doses of cattle manure in relation to sunflower productive performance in Agreste Paraibano. *Tecnolologia & Ciência Agropecuária*, João Pessoa, 7 (2): 20-28.

Santos, L. G., Melo, F. V. S. T., Souza, U. O et al. (2010). Phosphorus and boron in sunflower grain and oil production. *Enciclopédia Biosfera*, Centro Científico Conhecer - Goiânia, 6 (11): 1-8.

Smiderle, O. J. and Lima, J. E. (2009). Productive performance of no-tillage sunflower cultivars in the Cerrado of Roraima. In: *6th Brazilian Congress of Oilseeds, Oils, Fats and Biodiesel*. Montes Claros, MG. 416-422.

Souto, P. C., Souto, J. S., Santos, R. V., et al. (2007). Litter decomposition and microbial activity in caatinga area. In: *Brazilian Congress of Soil Science*, 31, 2007, Gramado. Anals... Brazilian Society of Soil Science. CD-ROM.

Souza, W. L., Ferrari, R. A., Scabio, A. et al. (2005). Sunflower oil and ethanol biodiesel. *Biomassa & Energia*, 2 (1): 1-5. <http:// florestasenergeticas.com/arquivos/p_biodiesel_etanol_25930.pdf>.

Tomich, T. R., Rodrigues, J. A. S., Gonçalves, L. C. et al. (2003). Forage potential of sunflower cultivars produced in off-season for ensiling. *Arquivo Brasileiro de Medicina Veterinária e Zootecnia*, 55 (6): 756-762.

Ungaro, M. R. G. (2000). *Sunflower Culture*. Campinas: Instituto Agronômico de São Paulo. (Boletim Técnico, 188).

United States Department for Agriculture - USDA. (2011). Oilseeds: World markerts and trade. Foreing Agricultural Service/USDA, *Circular Series*, September.

In: Sunflowers
Editor: Érico de Sá Petit Lobão

ISBN: 978-1-53617-195-2

Chapter 3

STAGES OF SUNFLOWER DEVELOPMENT: CORRELATIONS IN MALE AND FEMALE REPRODUCTIVE STRUCTURES FORMATION

Olga N. Voronova *
Komarov Botanical Institute,
Laboratory Embryology and Reproductive Biology,
Saint-Petersburg, Russia

ABSTRACT

Schneiter and Miller (1981) offered description of sunflower growth stages highlighting several vegetative stages (V1, V2... etc) and reproductive Stages (R1, R2...). The classification was based on external morphological features. The selection of several reproductive stages (R5.1, R5.2 etc) is based on the percentage ratio of blossoming flowers and does not take into account the development of male and female reproductive structures. Toderich (1988) investigated the correlations between male and female reproductive structures development. She identified 8 stages in the development of a tubular flower and described each in details.

Using our own data on the development of male and female reproductive structures in sunflower, we propose to combine and improve

* Corresponding Author's Email: o_voronova@binran.ru.

their classifications. We match the number of flowering circles in the head with changes in the reproductive structures - the phases of pollen formation and ovule development. Thus, the external morphological characteristics of the head (inflorescences), as a number of flowering circle, are associated with the peculiarities of the development of the internal structures of the flower. In our version of the classification, a simple and understandable division into the flowering stage of the head with a description of the state of the reproductive organs of the flower and its photographs is proposed.

Keywords: *Helianthus annuus*, sunflower development, reproductive system, ovule, embryo, anther, pollen

INTRODUCTION

Sunflower is one of the major world oilseeds. Sunflower breeding is very important for agriculture.

Sunflower plants pass through several development stages from planting to dieback. The most important developmental stages are a vegetative phase and a reproductive phase. Knowledge of the characteristics of the sunflower development and the correlations between the male and female reproductive sphere formation is necessary for successful breeding work.

The early embryological studies of cultivated sunflower were carried out by Goldflus (1899) and Nawashin (1900a,b) who described different aspects of reproductive system development. Most studies of sunflower reported that the mature pollen is three-celled and the ovule is tenuinucellar, unitegmic, anatropous with integumentary tapetum, *Polygonum*-type development of embryo sac, and *Asterad*-type (*Senecio*-variation) embryo formation (Newcomb 1973a,b; Toderich 1988; Yan et al. 1991; Gotelli et al. 2008; Voronova 2010; 2014). However, in all these works the connection between the formation of reproductive structures and the development phases of a plant in total was not considered. In the same time, it is useful and in some cases necessary to understand and to identify when the plant is at, or has moved through different growth and development stages.

A standardized and easy classification of sunflower development could help farmers, agronomist, scientists and other interested parties, clearly and accurately describe different sunflower growth stages.

METHODS

The sunflower plants of *Helianthus annus* cv. Peredovik and lines BK 571, VIR 114, VIR 116, VIR 151 were the main materials of the study. Plants were grown in the Kuban experimental station of Federal Research Center N. I. Vavilov All-Russian Institute of Plant Genetic Resources (VIR) in Krasnodar region (45°12′54.95″N, 40°47′37.12″E).

Plants were cross-pollinated. The whole heads or separated tubular (disk) flowerets were collected and preserved at different stages of flower development, starting from the ovule initiation stage (about 1cm heads diameter) to the torpedo embryo stage (over 20cm heads diameter). For the study, we took tubular flowers located, as a rule, on the edge of the head, mainly 1 - 2 flowering circles.

The materials were preserved in FAA solution (formalin, acetic acid and 70% ethanol in 7:7:100 ratio). The pieces were then dehydrated in an ethanol series, infiltrated in ethanol-chloroform mixtures and embedded in Histomix®. Sections for light microscopy (4 - 10mkm) were stained with various dyes in some combinations: toluidine blue; Ehrlich's haematoxylin with eosin or alcyan blue; Heidenhain's iron haematoxyline with alcyan blue (Zhinkina and Voronova 2000).

The sections were observed and photographed using Zeiss Axioplan 2 Imaging microscope. Schematic drawings are made on the basis of microphotographs. Schematic drawings help to better represent the real arrangement of flower structures, since they are not all and not always presented in the microphotographs correctly. This is due to the fact that the sunflower flowers are rather elongated and slightly curved towards the center of the head and therefore it is difficult to obtain a perfect longitudinal section.

In the tradition of plant embryological research, when representing reproductive structures, close-ups picture of ovule with macrosporocyte (macrospores tetrad, embryo sac) are placed with a micropylar area upward, and a chalasal region downward. We use the same principle in our illustrations and specifically mention this to avoid confusion, since the ovule in sunflower curves and turns 180 degrees in the process of development. Therefore, in general photographs and drawings, flowers (and ovules) are oriented as they are located in a head, in more detailed ones - with a micropylar area upward.

RESULTS AND DISCUSSION

Schneiter A. A. and Miller J. F. (1981), as shown in table 1, offered description of sunflower growth stages highlighting several Vegetative stages (V1, V2... etc) and Reproductive Stages (R1, R2...). The classification was based on external morphological features. The selection of several reproductive stages (R5.1, R5.2 etc) is based on the percentage ratio of blossoming flowers and does not take into account the development of male and female reproductive structures.

The growth stage key is divided into either Vegetative (V) or Reproductive (R) stages of plant development.

Vegetative development is further divided into two phases, Vegetative Emergence (VE) and true leaf development.

Vegetative Emergence (VE) covers the period from seedling emergence to when the first true leaf is less than 4cm long.

The next stage is the Vegetative (V) period with the growth stage given as "V" plus the number of true leaves over 4cm in length. For example, if there are two leaves over 4cm the growth stage would be V2, if there are four leaves it would be V4 and so on. If lower leaves have died and fallen off, one counts the leaf scar and include in the assessment.

The next stage is Reproductive (R) and it is separated into nine stages based on flower development. R1 to R4 describe the stages from when the flower bud first emerges through to just before the start of flowering. R5 describes the beginning of flowering and is sub divided to describe the percentage of the flowering that has completed or is in bloom such as R5.1 (10%), R5.5 (50%), R5.9 (90%). R6 to R9 cover the period form when flowering is completed (R6) through to physiological maturity (R9) and harvest.

The total time required for the development of sunflower plant and the time between the various stages of development depends on the genetic background of the plant and the growing environment. When determining the growth stage of sunflower field, the average development of a large number of plants should be considered. This staging method also can be used for individual plants. The same system can be used for classifying a single head or branched sunflower. In the case of branched sunflower, determinations make using only the main branch or head. At stages R7 through R9, healthy, disease-free heads were use to determine plant development if possible because some diseases can cause head discoloration.

Later (Figure 1), Schneiter et al. (2013) supplemented this scheme with colored photos and made it more visual.

In this scheme, the entire sequence of reproductive stages is described only by the external view of the sunflower head. The internal state and stages of development of the reproductive system are not taken into account.

Toderich (1988) studied the sunflower embryology, investigated the correlations between male and female reproductive structures development, and described it in her PhD thesis (Toderich 1988). Unfortunately, her work exists only in the form of manuscript and has not been published, so it is practically inaccessible for reading and her classification has not found practical application.

In the genesis of the tubular flower of sunflower (Table 2), taking into account the development of the male and female sphere, the author identified 8 stages using of *H.annuus* as example. The disadvantage of this classification is in the connection with a single tubular flower. Because that it is difficult to classify the reproductive status of the whole plant. As it is known, sunflower flowering occur centripetally, and within the same head, there are several rows of flowers at the different stages of development.

Table 1. Sunflower growth stages and description (Schneiter and Miller 1981)

Stage	Description
VE Vegetative Emergence	Seedling has emerged and the first leaf beyond the cotyledons is less than 4cm long
V (number) Vegetative Stages (e.g., V-1, V-2, V-3 etc.)	Determined by counting the number of true leaves at least 4cm in length beginning as V-1, V-2, V-3 etc. If senescence of the lower leaves has occurred, count leaf scars (excluding cotyledons).
R1 Reproductive Stages	The terminal bud forms a miniature floral head rather than a cluster of leaves. When viewed from directly above, the immature bracts have a many-pointed star-like appearance.
R2	The immature bud elongates 0.5 to 2.0cm above the nearest leaf attached to the stem. Disregard leaves attached directly to the back of the bud.
R3	The immature bud elongates more than 2cm above the nearest leaf.
R4	The inflorescence begins to open. When viewed from directly above, immature ray flowers are visible.
R5 (number) e.g., R-5.1, R-5.5, R-5.9, etc.	The beginning of flowering, divided into substages depending on the percent of the flower that has completed or is in flower, eg. R-5.1 (10%), R-5.5 (50%), R5.9 (90%).
R6	Flowering complete, ray flowers wilting.
R7	Back of the head started to turn a pale yellow.
R8	Back of the head yellow, bracts remain green.
R9	Bracts yellow and brown, plant at physiological maturity.

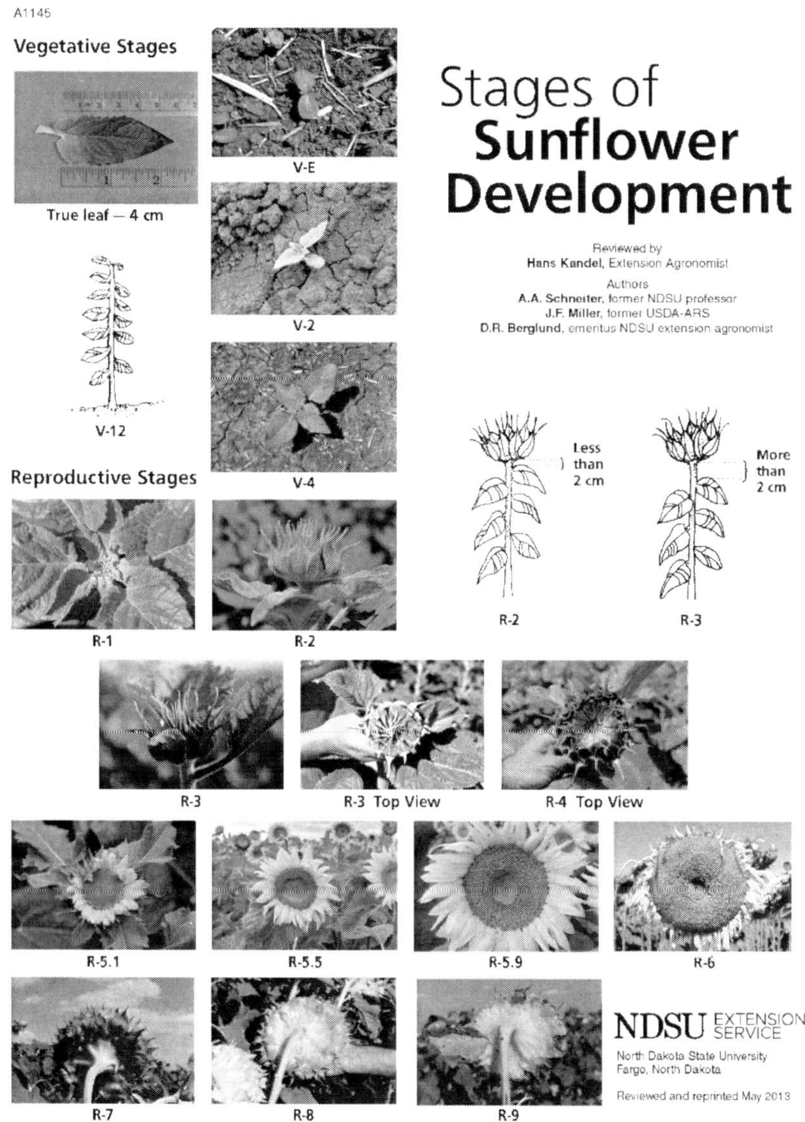

Figure 1. Stages of sunflower development (Schneiter et al. 2013).

Another disadvantage is that it almost not illustrated. Only one small table with fuzzy photos accompanies the description of the stages. Thus, this attempt to determine the stages of development of an individual flower is interesting from a scientific point of view, but cannot be used in practice.

Table 2. Tubular flower growth stages and description (Toderich 1988)

Stage	Description
1.	Closed bud, stamen sessile, located at the base of the corolla (development of the anther wall), the formation of the ovary cavity.
2.	The bud is still closed, the style is immature, the stamens reach 1/3 the length of the corolla (anther wall formed, meiosis), the beginning of ovule development (differentiation of megasporocyte).
3.	The beginning of disclosure the bud, the corolla is significantly lengthened; by the end of the stage the stamens reach more than a half of its length (in anthers tetrad of microspores, tapetum begins the regeneration in periplasmodia, division in microspore), the style increases in size significantly (in the ovule meiosis and the embryo sac development are observed).
4.	The flower is open. The corolla reaches almost the maximum size; the stamens and the style reach the level of the corolla; the stamen filaments grow considerably, the anthers on the eve of disclosure (division of the generative cell and maturation of sperm cells). In the ovule there is a differentiation of the elements of the embryo sac.
5.	Fully open flower, the corolla stops growth, their filaments straightened and reached its maximum development; the anthers are completely cracked (mature pollen). The closed lobes of the stigma, with adhering pollen, protrude from the anther tube. (In the ovule there is a mature embryo sac with differentiated cells).
6.	Fully open flower. The anthers are wrinkled, they filaments twist spirally, and the stigma's lobes acquire a horizontal position. Ovary begins to lengthen (fertilization is completed).
7.	Fading flower. The stigma lobes twist spirally and at the end of this stage are drawn inside the corolla. Enhanced growth of the ovary. (Development of endosperm, embryo).
8.	The stage of seed formation and ripening. The corolla together with the anther tube and the faded style falls off, or remains attached to the ovary. Ovary reaches its maximum development.

Cetinbas and Unal (2012) made a rather detailed and well-illustrated description of the development of a tubular flower. However, their work was devoted to a comparison of tubular (hermaphrodite) and ray (pistillate) flowers and they made no attempt to compare the developmental phases of a single flower with the flowering of the whole head or to distinguish the development process into separate stages.

We studied in detail the early stages of development of the female reproductive sphere in cultured sunflower (Voronova 2014), as well as some aspects of embryology in several wild species (Voronova and Babro 2018; Babro and Voronova 2018) and some forms of apomixis in sunflower (Voronova 2010; Voronova and Babro 2019).

Using our data on the development of male and female reproductive structures in sunflower and taking into account the research Toderich (1988) we propose to improve the classification of Schneiter and Miller (1981) and to make some changes in it (Table 3).

We propose to change the division into substages in the R-5 stage. Instead of the previously proposed counting the percentage of flowers in bloom from the total number of flowers, we suggest taking into account the number of flowering circles. Tubular flowers in a sunflower head bloom not all at the same time, but gradually circle by circle. Flowering spreads centripetally. The width of the circle, on average, ranges from 1 to 2 cm and depends on the size of the head and individual flowers in it. According to the sequence number of the flowering circle, we propose to call substages in the R-5 stage. Thus, if the first circle blooms, it will be R-5.1, if the second circle - R-5.2 and so on.

Description and assessment of the development of reproductive structures was carried out on tubular flowers located on the edge of the head, as the most advanced in development. Moving from the edge of the head to the center, one can observe earlier stages of flowers development.

Taking into account everything said above, stages R-1 to R-5 will have the following additional descriptions and photographs.

The main thing that draws attention to when analyzing the resulting table is the asynchrony of the laying and development of similar structures in the male and female sphere.

Table 3. Correlations in the sunflower male and female reproductive structures formation

Reproductive Stage - R-1 The terminal bud forms a miniature floral head rather than a cluster of leaves. When viewed from directly above, the immature bracts have a many-pointed star-like appearance.
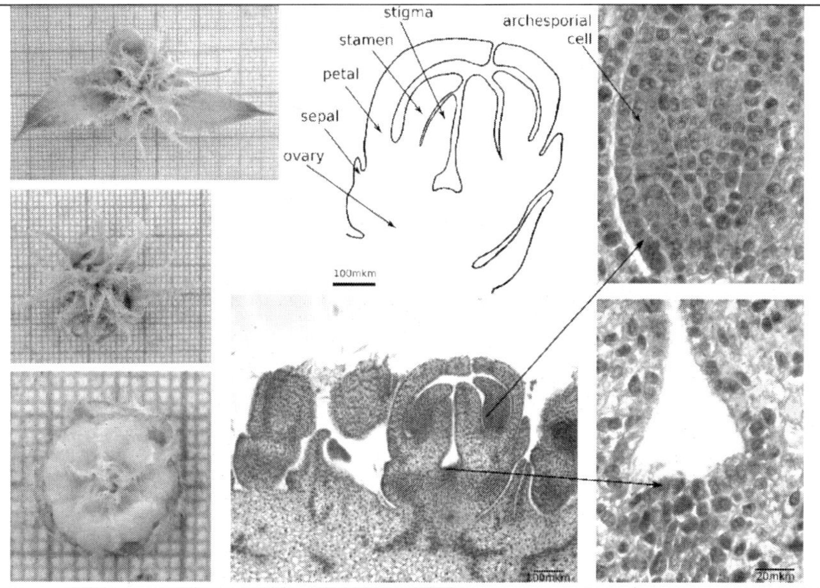
Flower primordium. Petals development and androecium differentiation. There is an epidermal layer and uniform cells inside it at the beginning of anther development. Anthers during the first division of archesporial cell. Gynoecium does not occur yet.

Reproductive Stage - R-2

The immature bud elongates 0.5 to 2.0cm above the nearest leaf attached to the stem. Disregard leaves attached directly to the back of the bud.

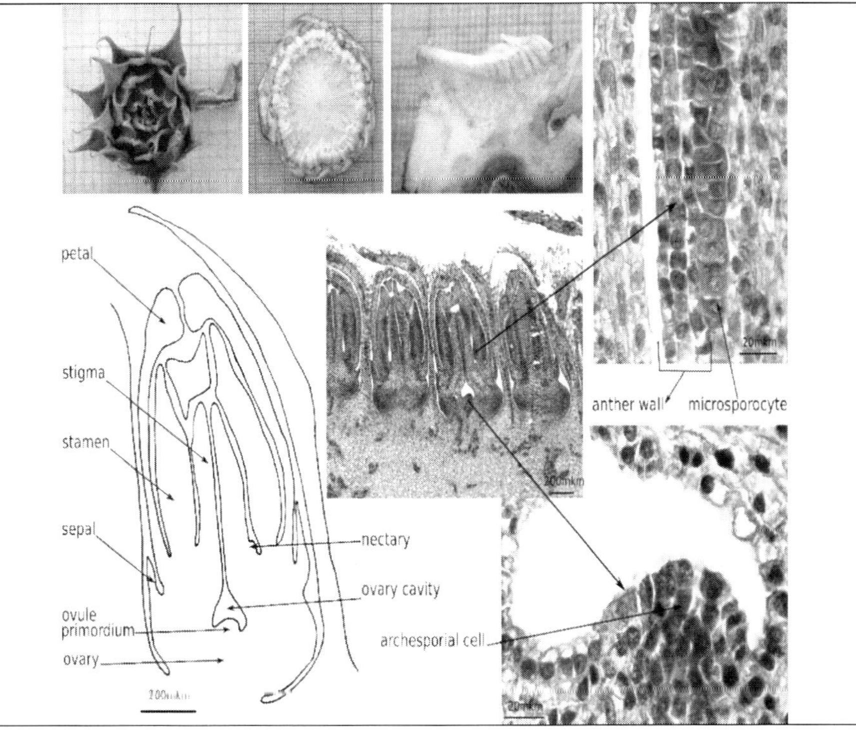

Closed bud, stamen sessile, located at the base of the corolla. Differentiation of wall layers in microsporangia. Anther wall develops centrifugally: tapetum differentiates at first, and later the endothecium and the middle layer are formed. The formed anther wall consists of four layers – epidermis, endothecium, middle layer and tapetum. The number of microsporocytes increases and they undergo meiosis. The ovary cavity forms. In the ovule primordium - the division of the subepidermal cells to form archesporial cells; the initial stages of the integument formation.

Table 3. (Continued)

Reproductive Stage - R-3 The immature bud elongates more than 2cm above the nearest leaf.
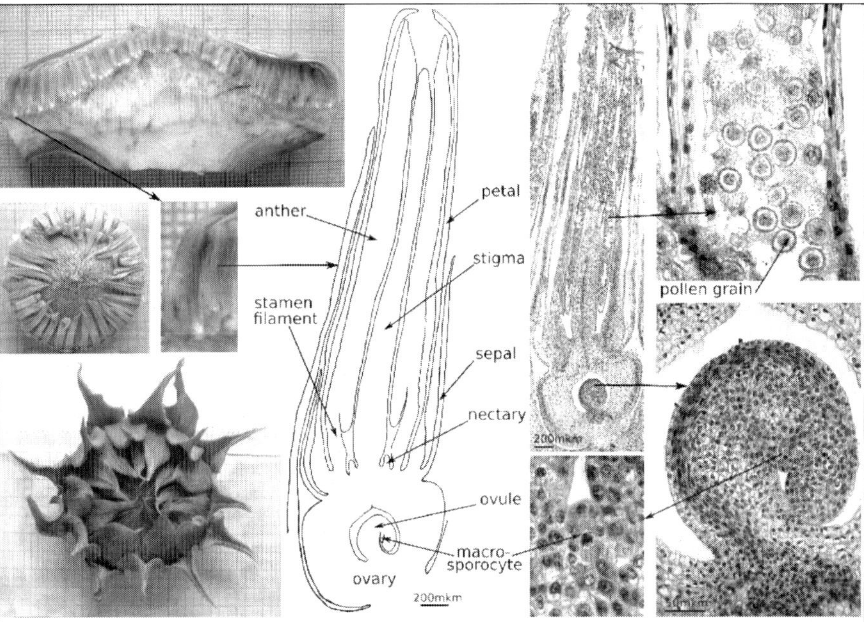
The beginning of disclosure the bud. The corolla is significantly lengthened; by the end of the stage the stamens reach more than a half of their length. In anthers tetrad of microspores pass to free one-celled pollen grains; tapetum cells begin the regeneration in periplasmodia. The style increases in size significantly. In the ovule microsporocyte which pass threw different stages of meiosis is observed.

Reproductive Stage - R-4

The inflorescence begins to open. When viewed from directly above, immature ray flowers are visible.

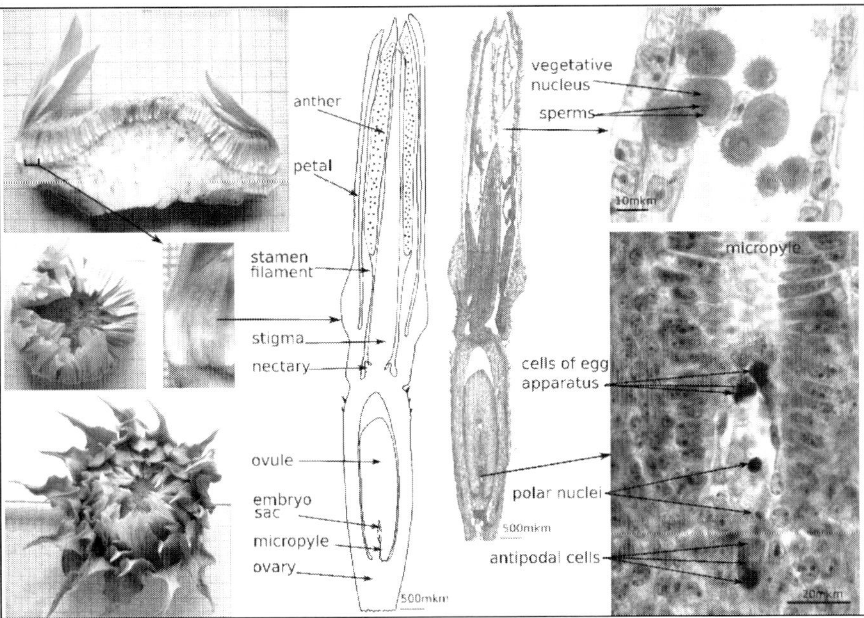

The flower is nearly open.
The corolla reaches almost the maximum size.
The stamens and the style reach the level of the corolla; the stamen filaments grow considerably, the anthers on the eve of disclosure (division of the generative cell and maturation of sperm cells).
In the ovule a differentiation of the elements of the embryo sac occurs. Embryo sac consists of three-celled egg apparatus, central cell with polar nuclei and antipodes (2 or 3 celled).

Table 3. (Continued)

Reproductive Stage - R-5

The beginning of flowering, divided into substages depending on the number of flowering circles.

R-5.1

The first circle of flowers blooms

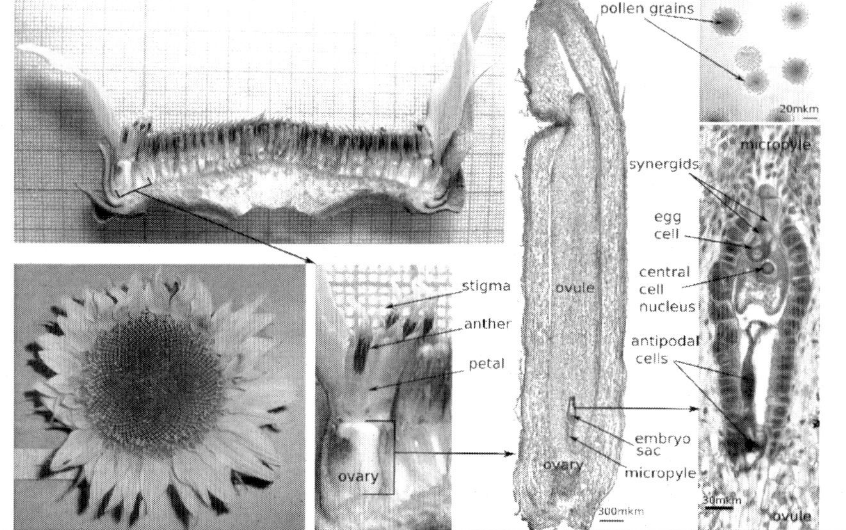

Fully open flower.
The corolla stops growth, the filaments straightened and reached its maximum development. The anthers are completely cracked. Mature pollen grains are moved to the top of the anthers from where they are carried away by insects. In the end of this stage, the closed lobes of the stigma protrude from the anther tube. In the ovule, there is a mature embryo sac with differentiated cells. Embryo sac consists of egg cell, two synergids, central cell with fused nucleus and antipodes (2 or 3 celled).

R-5.2 The second circle of flowers blooms.
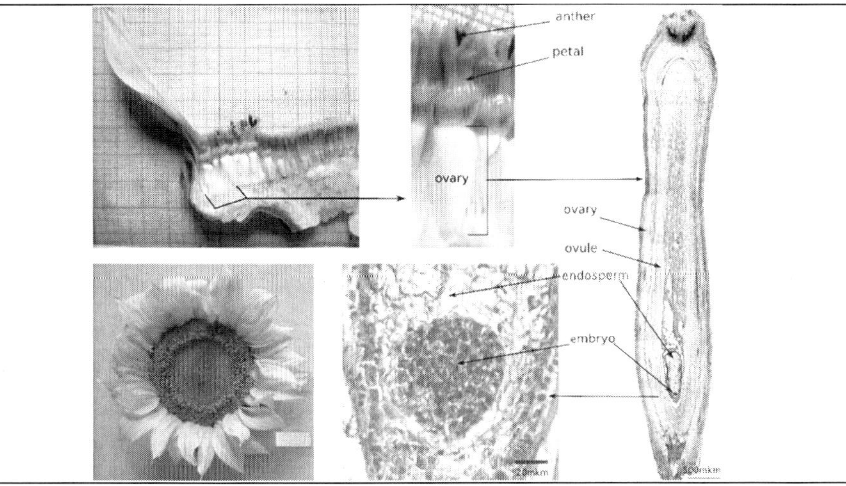
Pollination and seed formation. The stigma lobes twist spirally and at the end of this stage are drawn inside the corolla. Pollination and fertilization is completed. In the ovule, there is a young embryo (globular stage) and endosperm.
R-5.3 The third circle of flowers blooms.
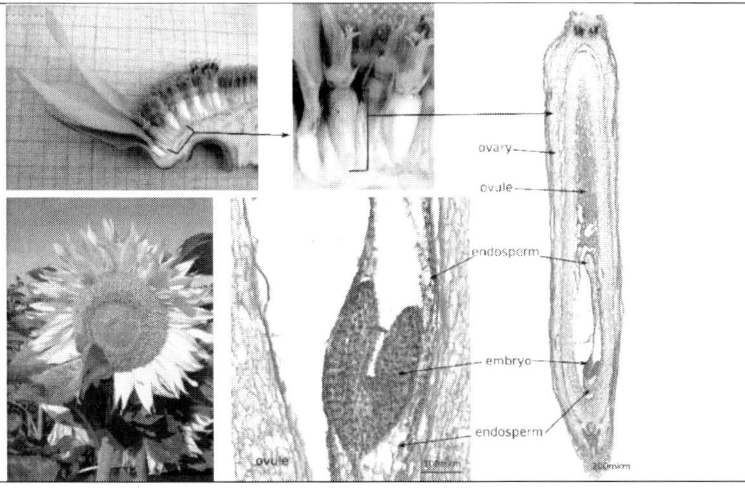

Table 3. (Continued)

R-5.4 The center of head blooms
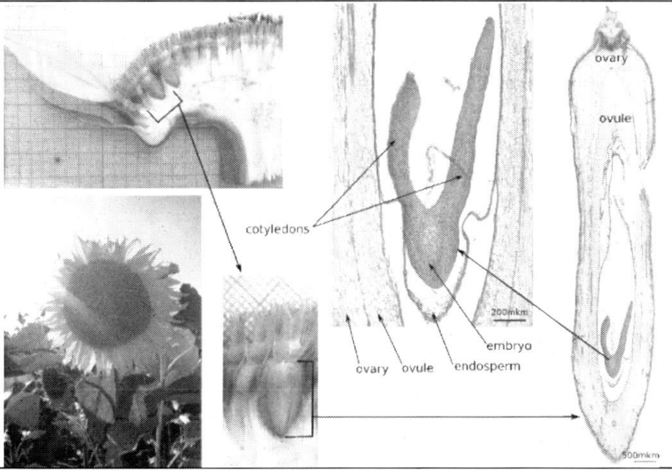
Seed formation and ripening. Ovary reaches its maximum development. In the ovule, there is a late torpedo-shaped embryo. Cotyledons grow intensively and fill the cavity of the ovule. Endosperm starts degenerated.

<table>
<tr><td>R-6</td><td>Flowering completes, ray flowers wilting.</td><td rowspan="4">The embryo is growing rapidly. The wall of the ovary becomes the seed coat, and the ovule tissue becomes thinner and remains in the mature seed as a film.</td></tr>
<tr><td>R-7</td><td>Back of the head started to turn a pale yellow.</td></tr>
<tr><td>R-8</td><td>Back of the head yellow, bracts remain green.</td></tr>
<tr><td>R-9</td><td>Bracts yellow and brown, plant at physiological maturity.</td></tr>
</table>

The development of male reproductive organs starts earlier than female ones. Archesporial cells in the anthers are laid on the stage R-1. Microsporocytes form and pass meiosis starting from stage R-2. Formation of microspores (pollen grains) begins at the stage R-3, to the stage R-4 in the anthers are already formed pollen grains, which are fully mature at the time of flowering (stage R-5).

Archesporial cells in the ovule are detected at the R-2 stage. They develop into macroporocytes, which at the stage of R-3 enter into meiosis, which leads to the formation of macrospores. During the R-4 stage, a female gametophyte, an embryo sac, is formed from the macrospore. The maturation of the female gametophyte occurs at the time of flowering - to stage R-5.

Thus, at the time of flowering, both the male (pollen grain) and the female (embryo sac) gametophytes are fully ripe and ready for pollination and fertilization. At the same time, male and female meiosis in the reproductive cells are separated in space and time. This is important to maintain the reliability of the reproduction system, since meiosis is a phase of the cell cycle during which cells are most sensitive to external influences, such as the effects of mutagenic substances.

In addition, knowledge of key points in the development of structures can help to achieve the impact of chemical reagents selectively on the male or female reproductive sphere.

For example, it was experimentally found that gibberelin treatment to produce sterile anthers is better carried out at the stage morphologically corresponding to P-1. It is clearly seen from the table that it is at this stage that the archesporial cells are laid in the anther, while the hyphae are not yet formed and are not significantly affected.

CONCLUSION

The proposed table gives a visual representation of the order of the developmental phases and the correlation between the male and female generative spheres.

Proper assessment of the development stages of sunflower will undoubtedly be useful and significant for agrotechnical and breeding work.

For example, treatment with gibberellin to produce plants with a sterile male sphere is most effective in R-1 stages (before microsporocytes formation).

Also, when treating sunflower plants with substances with possible mutagenic properties, it is necessary to take into account that the cells are most sensitive to them during meiosis, which undergoes in sunflower in the female and male spheres at different times.

ACKNOWLEDGMENTS

I am grateful to Dr. V. A. Gavrilova (N. I. Vavilov Research Institute of Plant industry), Dr. V. T. Rozhkova, and Dr. T. T. Tolstaja (the Kuban experimental station) for their help in obtaining the experimental material.

The present study was carried out within the framework of the institutional research project (AAAA-A18-118051590112-8 "Polyvariance of morphogenetic programs for the development of plant reproductive structures, natural and artificial models of their implementation") of the Komarov Botanical Institute of the Russian Academy of Sciences.

REFERENCES

Babro, A. A. and Voronova, O. N. (2018). “Development of male reproductive structures in *Helianthus ciliaris* and *H. tuberosus* (Asteraceae)” *Bot. Zhurn.*, 103(9): 1093 - 1108. [in Russian]. https://doi.org/10.7868/ S0006813618090028.

Cetinbas, A. and Unal, M. (2012). “Comparative Ontogeny of Hermaphrodite and Pistillate Florets in *Helianthus annuus* L. (Asteraceae)”. *Not. Sci. Biol.*, 4 (2): 30 - 40.

Goldflus, M. (1899). “Sur la structure et les fonctions de l'assise épithéliale et des antipodes” [On the structure and functions of the epithelial base and the antipodes]. *J. de bot.*, 13: 87 - 96.

Gotelli, M. M., Galati, B. G. and Medan, D. (2008). "Embryology of *Helianthus annuus* (Asteraceae)". *Ann. Bot. Fennici,* 45 (2): 81 - 96.

Navashin, S. G. (1900a). "On fertilization in the Asteraceae and Orchids". *Proc. Imperial Acad. Science,* 13 (3): 335 - 340 [in Russian].

Navashin, S. (1900b). "Ueber die Befruchtungsvorgänge bei einigen Dikotyledonen" [About fertilization in some dicotyledons]. *Berichte der deutschen botanischen Gesellschaft,* 18 (5): 244 - 250.

Newcomb, W. (1973a). "The development of the embryo sac of sunflower *Helianthus annuus* before fertilization". *Can. J. Bot.,* 51: 863 - 878.

Newcomb, W. (1973b). "The development of the embryo sac of sunflower *Helianthus annuus* after fertilization". *Can. J. Bot.*, 51: 879 - 890.

Schneiter, A. and Miller, J. F. (1981). "Description of Sunflower Growth Stages". *Crop Sci.*, 21 (6): 901 - 903. https://doi.org/10.2135/cropsci1981.0011183X002100060024x.

Schneiter, A. A., Miller, J. F. and Berglund, D. R. (2013). "*Stages of Sunflower Development*". Reviewed by Hans Kandel, North Dakota State University. https://www.ag.ndsu.edu/publications/crops/stages-of-sunflower-development.

Toderich, K. N. (1988). "*Sunflower (Helianthus annuus, H. rigidus, etc.) embryology*". PhD diss., Komarov Botanical Institute, Leningrad, [in Russian].

Voronova, O. N. (2010). "Integumentary embryony in CMS sunflower line". *Russ. Journ. Dev. Biol.,* 41(6): 394 - 399. https://doi.org/10.1134/S106236041006007X.

Voronova, O. N. (2014). "Chapter 3. Development of female reproductive structures and apomixis in sunflower" in *Sunflowers: Growth and Development, Environmental Influences and Pests/Diseases*. New York: NOVA, 43 - 60.

Voronova, O. N. and Babro, A. A. (2018). "Early stages of formation of female reproductive structures in *Helianthus ciliaris* and *H. tuberosus* (Asteraceae)". *Bot. Zhurn.,* 103 (4): 488 - 504. [in Russian] https://doi.org/10.1134/S0006813618040051.

Yan, H., Yang H. Y. and Jensen, W. A. (1991). "Ultrastructure of the developing embryo sac of sunflower (*Helianthus annuus*) before and after fertilization". *Can. J. Bot.,* 69 (1): 191 - 202.

Zhinkina, N. A. and Voronova, O. N. (2000). "On staining technique of embryological slides". *Bot. Zhurn.,* 85 (6): 168 - 171. [in Russian].

In: Sunflowers
Editor: Érico de Sá Petit Lobão
ISBN: 978-1-53617-195-2

Chapter 4

CLASSIFICATION OF SUNFLOWER UNDER WATER STRESS CONDITIONS BY MEANS OF SPECTRAL REFLECTANCE AND CHEMOMETRICS

***Antonio José Steidle Neto**[*]*
and Daniela de Carvalho Lopes
Campus Sete Lagoas, Federal University of São João Del-Rei,
Sete Lagoas, Minas Gerais, Brazil

ABSTRACT

Sunflower is one of the economically important crops in the world. It is largely cultivated for biofuel production, represents a relevant energy source in human diet and is widely used as ornamental plant. However, the water stress is a limiting factor in sunflower cultivation, leading to an emphasis on developing innovative technologies that make it possible to maximize the water use efficiency, as well as to improve irrigation and management strategies. In this context, Factorial Discriminant Analysis (FDA) and Partial Least Squares Discriminant Analysis (PLS-DA) were used to classify intact sunflower plants under three water stress levels

* Corresponding Author's Email: antonio@ufsj.edu.br.

(without, moderate, and severe) based on Vis/NIR spectral reflectance measurements (500 - 1000 nm range). The raw and pre-treated (detrend, first and second derivatives) spectral signatures were centered and normalized before data processing. Results showed a satisfying performance when using both multivariate methods with raw and pre-treated spectra. But, PLS-DA using the first derivative pre-treatment performed slight better for classifying the sunflower according to different water stress conditions, reaching overall correct classification of 80.67% when considering the external validation dataset. The lower accuracy was verified for the FDA with second derivative, resulting in an overall correct classification of 69.05%. The Vis/NIR spectroscopy combined with chemometrics is a promising tool for non-destructive and fast identification of sunflower water stress and can be used into processing lines, enabling large-scale individual analysis and real-time decision making.

INTRODUCTION

Sunflower (*Helianthus annuus* L.) is one of the economically important crops for production of oil and forage in the world, combining high seed yields and great adaptability to a wide range of geographical areas (Patanè, Cosentino and Anastasi 2017). It is largely cultivated for biofuel production due to the oxidative stability of its oil with high oleic acid content (Polin et al. 2014). Additionally, it represents an important energy source in human diet and is a relevant ornamental plant, since sunflowers present attractive foliage and large inflorescence (Best et al. 2017).

However, water stress is a common limiting factor in sunflower cultivation and the worldwide water crisis has contributed to the increase of this problem in many parts of the world (Buriro et al. 2015; Steidle Neto, Lopes and Borges Júnior 2017a; Liu et al. 2018). This is leading to an emphasis on developing innovative technologies that make it possible to maximize the water use efficiency and to improve the productivity of crops. Among them, the assessment of plant water status under drought conditions has shown to be very important in terms of developing irrigation, selection and breeding strategies (Steidle Neto et al. 2017b).

The most reliable and accurate method to obtain plant water status is by determining mass differences between wet and dry leaves in laboratory, which is a sample-destructive, labor-demanding, and time-consuming method. But, several studies (Gamon and Surfus 1999; Steidle Neto et al. 2017c; Monazzah et al. 2018) indicated that rapid assessment of pigments, chemical properties, and water content in intact sunflower plants is possible using the combination of Vis/NIR spectroscopy and chemometrics. This approach has considerable advantages over traditional methodologies including cost, throughput, non-destructive sample preparation and analysis, as well as risk reduction where potentially dangerous chemicals and/or procedures are involved (Su, He and Sun 2017).

The spectroscopic techniques relate electromagnetic radiation to product properties at different wavelengths and can be applied in visible (VIS: 400 to 700 nm), near infrared (NIR: 700 to 2500 nm), mid-infrared (MIR: 2500 to 25000 nm), and far-infrared (FIR: 25000 to 1000 x 10^3 nm) spectra (Lopes and Steidle Neto 2018). On the other hand, chemometrics is a well-established science, which allows getting maximum information from spectral data by using multivariate statistical methods for the analysis of complex spectra. According to Diago et al. (2013), chemometrics combine the reduction of the spectroscopic data dimension and the creation of models that transform independent variables (wavelengths) into dependent ones (plant or food characteristics), improving and speeding up the analysis of the relationship between the product spectra and properties.

Despite the potential of this technology and the increase in pressure due to the water resources and atmospheric precipitation reduction in several world regions, published studies for the responses of sunflower crops under drought conditions based on reflectance measurements together with multivariate data analysis are scarce (Steidle Neto et al. 2017b; Steidle Neto et al. 2017c). The existing researches have focused on quantification of water content, and there is a lack of classification models capable of differentiating between sunflower leaves without water stress and those under moderate and severe conditions.

The automatic classification of plants is increasing in importance, since there is a need to optimize the on-line monitoring procedures in field and industry, improving remote sensing identification and assuring product credibility. Specifically, the separation of distinct sets of intact sunflower plants into different categories of water stress contributes to the management techniques and water use efficiency by plants (Chaves, Flexas and Pinheiro 2009).

There are several applications that confirm the cost-effectiveness of spectroscopy combined with chemometrics for automatic classification of plants (Berrueta, Alonso-Salces and Héberger 2007; Moros 2010; Symonds et al. 2015, Steidle Neto et al. 2018). Further, the evolution of electronics and the expansion of the agro-food industry have enabled access to technologies that before were only available in well-equipped laboratories and research centers. However, there are still challenges to introduce effective techniques for this purpose, since different plants and research objectives imply in different responses when the discriminant methods and spectral pre-treatments are applied (Sánchez et al. 2013).

In this context, the present study aimed at classifying sunflower plants based on different water stress conditions, using Vis/NIR spectral reflectance measurements and multivariate statistical methods (FDA and PLS-DA) applied to raw and pre-treated (detrend, first and second derivatives) spectra, which were centered and normalized before data processing.

MATERIAL AND METHODS

Cultivation

Sunflower plants of variety Sunbright (Sakata Ornamentals, Morgan Hill, California, USA) were cultivated inside a non-acclimatized greenhouse located at the experimental area of the Agrarian Sciences Department of the Federal University of São João del-Rei, Sete Lagoas (city), Minas Gerais (state), Brazil (19°28' S latitude, 44°11' W longitude, and 796 m elevation). According to Köppen's classification, the local climate is humid subtropical with dry winter and hot summer (Alvares et al. 2013). The greenhouse was oriented with the ridge running east to west, built as a Quonset frame with galvanized structural steel tubing, and covered with a transparent polyethylene film (150 μm).

The hybrid Sunbright was chosen for this study because it is very popular with consumers and growers since it is standard as cut flower due to its bright golden petals and great shipping ability. This sunflower variety is pollenless, very uniform, and vigorous in growth even under unfavorable weather conditions. The total number of sunflower plants cultivated in the greenhouse was greater than the number required for the spectral reflectance measurements, because only the healthy, vigorous, and well-formed plants were selected.

Sunflower cultivation was carried out during the spring seasons of 2014 and 2015, following suitable agronomic practices of management, fertilization, and irrigation. This experimental design was established in order to account for the inter-annual variability of climatic variables (duration of sunshine, air temperature, and air relative humidity) that can influence plant growth and development. The water and nutritional conditions under the greenhouse cultivation were identical for both experiments and the information collected were joined for the chemometric analysis, searching for a more representative dataset capable of improving the robustness of the developed models.

Sunflower seeds were germinated in commercial substrate consisting of pine bark, vermiculite, and peat (4:3:3 v v^{-1}), with mean density of 0.42 g cm^{-3} and contained in plastic pots with capacity of 905 cm^3. The irrigation and fertigation events were automatically supplied to plants by a time-controlled drip system connected to a 37.3 W pump and a 220 L water reservoir. The control strategy was configured so that irrigations and fertigations were performed in alternate days (100 mL $event^{-1}$ $plant^{-1}$), providing a proper nutrient balance in the substrate and preventing root salinity (Steidle Neto, Lopes and Borges Júnior 2017a). Nutrient solution of 1 dS m^{-1} was prepared from granulated fertilizer (15 N - 5 P - 15 K - 5 Ca - 2 Mg, Peters Excel, Scotts, Marysville, Ohio, USA), containing 150 mg of nitrogen per liter. Electrical conductivity cells connected to an electronic circuit (Steidle Neto et al. 2005) were used for monitoring the applied and leachated nutrient solution.

Four applications of a growth regulator (Pachlobutrazol 100 CE, Wiser, São Paulo, São Paulo, Brazil) were performed on substrate nearest of each plant stem aiming at avoiding plant tutoring and making easy the plant management at laboratory. Each dose contained 4.5 mg $plant^{-1}$ of active ingredient and was applied at weekly intervals starting from the second week after seeding. This growing control technique resulted in sunflower plants with average height of approximately 0.4 m.

Approximately 60 days after seeding, at the start of the sunflower flowering stage, the last irrigation event was accomplished at early morning with the objective of maximizing the substrate moisture. After this day, no fertigation events were performed, and sunflower plants were transported to a reach-in grown chamber where they were submitted to a slow and progressive dehydration rate for 12 consecutive days. The chamber was equipped with actuators (heating, cooling, fogging, and lighting systems) and sensors (photometric, pyranometer, air temperature, and air relative humidity), and was programmed to control internal conditions for 1600 fc (foot-candle, 17.2 klux) illumination, 12 h photoperiod, 30 to 25°C (day–night) temperature, and 40% humidity.

Spectrometric and Laboratorial Measurements

Spectral reflectance and leaf mass (fresh and dry) were made in three plants of each experiment, randomly selected once a day.

The reflectance measurements were accomplished in three leaves of each plant, which were randomly selected from the upper and middle thirds of the sunflower stem, since that many leaves located in the lower third were senescent when the plants were transferred to the grown chamber. Three separate measurements on adaxial surface were performed in each leaf, avoiding its central vein and boundaries. For each leaf, average spectral reflectance curves were generated to analyze these measurements, totaling 648 spectral signatures (12 days x 3 plants x 3 leaves x 3 points per leaf x 2 experiments).

The spectral signatures were obtained by using a miniature spectrometer (JAZ-EL350, Ocean Optics, Dunedin, Florida, USA) coupled to a tungsten-halogen light source. Reflectance data in Vis/NIR wavelength range (500 - 1000 nm) were acquired and stored with a spectral resolution of 1.3 nm. A reflection probe (R400-7-VIS-NIR, Ocean Optics, Dunedin, Florida, USA) was used to collect reflected light from the sunflower leaves and was connected to the spectrometer and light source, respectively. This probe is a bifurcated optical fiber assembly (Y type) composed of two fibers of same diameter (400 µm). The other probe tip was inserted in a holder of anodized aluminum, being vertically positioned in relation to leaves for diffuse reflectance measurements (Figure 1).

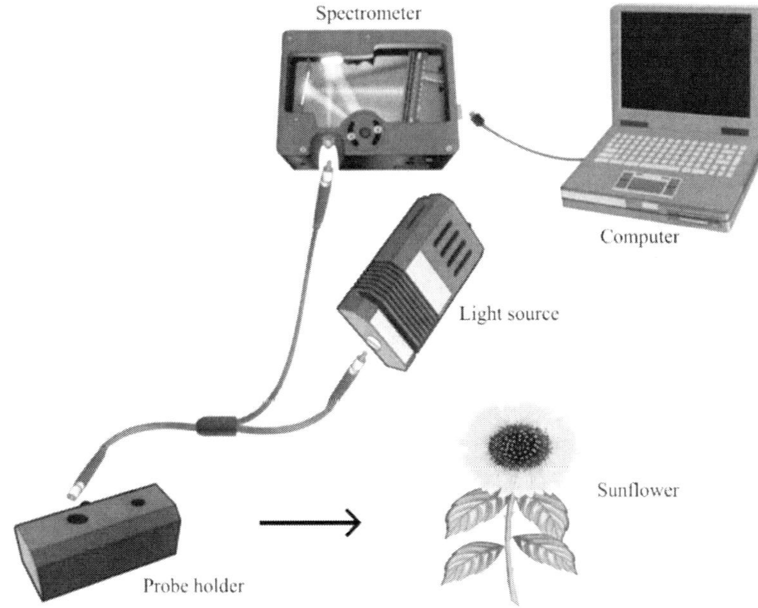

Figure 1. Scheme of the equipment used for measuring the spectral reflectance of sunflower.

Daily, the light source was warmed up first (10 min) and reference standard measurements were made before any spectral reflectance measurements from sunflower leaves with the purpose of calibration. A diffuse reflectance standard (WS-1-SL, Ocean Optics, Dunedin, Florida,

USA) with Spectralon™ was used as a reference to measure spectral reflectance, while the spectral reflectance considering light absence was obtained by obstructing the light input at the holder.

Reflectance values were calibrated by software (OceanView, Ocean Optics, Dunedin, Florida, USA) and expressed as a relative percentage of the reference standard:

$$R_{\lambda}^{cal} = 100\ [(R_{\lambda}^{leaf} - R_{\lambda}^{dark}) / (R_{\lambda}^{ref} - R_{\lambda}^{dark})] \tag{1}$$

where R_{λ}^{cal} is the calibrated spectral reflectance from the leaves (%), R_{λ}^{leaf} is the original spectral reflectance from the leaves (dimensionless), R_{λ}^{dark} is the spectral reflectance considering light absence (dimensionless), and R_{λ}^{ref} is the spectral reflectance from the reflectance standard (dimensionless).

After the non-destructive measurements, each leaf was cut at the base of the petiole and its fresh mass was obtained on an analytical balance (AL-500, Marte, São Paulo, São Paulo, Brazil). The leaves were then individually placed in paper bags and dried at 70°C in an oven (MA-035/1, Marconi, São Paulo, São Paulo, Brazil) until a constant mass. Again, each leaf was weighed, and its dry mass was obtained. The reference leaf water content was calculated dividing the difference of fresh and dry masses by the dry mass (Ullah et al. 2013):

$$L_{W} = (L_{m}^{fresh} - L_{m}^{dry}) / L_{m}^{dry} \tag{2}$$

where L_W is the leaf water content (dry basis), L_m^{fresh} is the leaf fresh mass (g), and L_m^{dry} is the leaf dry mass (g).

The reference leaf water contents were required in order to calibrate the chemometric models. Thus, each reference value was associated with a spectral signature of sunflower and the dataset was separated into three water stress levels (without, moderate, and severe) based on the measured leaf water contents.

Chemometric Analysis

Factorial Discriminant Analysis (FDA) and Partial Least Squares Discriminant Analysis (PLS-DA) were used to evaluate the potential of Vis/NIR spectroscopy to discriminate three water stress levels (without, moderate, and severe) in sunflower plants. For this, the whole raw spectra were considered, as well as the effects of three spectral pre-treatments (detrend, first and second derivatives) on the classification were also tested. Further, all spectra were centered and normalized before each data processing.

Data pre-treatment can considerably affect the results of discriminations based on spectral data. Frequently, spectral signature does not visibly differ among themselves, also containing undesirable components, called noise, which can reduce the computing efficiency of the proposed models (Lopes and Steidle Neto 2018). Spectral pre-treatments can help to remove irrelevant information which cannot be handled by the chemometric methods, enhancing spectral differences between classes. On the other hand, the excessive pre-treatment or the inadequate use of these techniques can result in loss of spectral information that can be useful for the classification of models (Moscetti et al. 2015). For this, the abovementioned preprocessing techniques were used and tested, improving the evaluation of the robustness and the accuracy of the developed models.

According to Steidle Neto et al. (2018), centering can improve the classification accuracy for most of the discriminant methods, by enhancing the differences between spectra. The normalization adjusted the spectral data from the different groups (water stress levels) to an identical baseline, facilitating subsequent spectral analysis and comparisons in discrimination purposes (Yuan et al. 2014). Centering and normalization were calculated following the procedures recommended by Martens and Naes (1992). For centering, an average spectrum was calculated by using all the spectra in the dataset and then it was subtracted from each spectrum (mean-centering). The Standard Normal Variate (SNV) was applied as normalization technique, in which the spectrum was subtracted at every wavelength by the spectrum average and this result was divided by the spectrum standard deviation.

Detrend is a baseline correction algorithm capable of correcting simple deformations of the spectra baseline as vertical shift and slope (Moura et al. 2016). This pre-treatment was obtained by an orthogonal projection of the raw spectra to a Vandermonde matrix, in which a polynomial fit of each spectrum was subtracted from it. Spectrum derivative allows removing baseline shifts, minimizing overlapped peaks and correcting both additive and multiplicative effects (Shao et al. 2015). First and second derivatives were calculated by using the Savitzky-Golay algorithm (Savitzky and Golay 1964) with 25 derivative points (window for calculation), optimally fitting the dataset points to a polynomial in the least-squares sense.

Both FDA and PLS-DA are supervised methods and use a training set of samples to define a decision boundary in the space of response patterns. Thus, data was grouped into predefined classes and mathematical models were built, capable of identifying unknown samples with different water stress levels without destroying the sunflower leaves.

A Principal Component Analysis (PCA) was performed before FDA and PLS-DA to derive the principal components (PC) from the spectral data, seeking inherent similarities of data, as well as detecting patterns and outliers in the dataset (Wang and Mizaikoff 2008). The first principal component represented the largest amount of variability in the original dataset, and each succeeding component accounted for as much of the remaining variability as possible. Thus, the original data was transformed into new variables (PCs) which were orthogonal and uncorrelated. The PCA was run with the spectral reflectance values following procedures proposed by Saporta (2006), where original data matrix was decomposed into score, loading, and residual matrices.

Loading matrix represented the correlation of the original variables with the PCs, while residuals meant the part of data that were not explained by the PCA method. Score matrix represented the coordinates of the transformed variables in the PC space. That is, it was related with the original regressors and was used for data exploration and model predictions.

When applying FDA and PLS-DA methods, the calibration with cross-validation followed by an external validation was used, since it is the most recommended procedure when developing chemometric models (Westad and Marini 2015). According to Lopes and Steidle Neto (2018), this procedure allows that the same samples have the probability to be used for training and testing the chemometric models based on equal pre-processing and multivariate techniques. Thus, dataset was divided in groups and models from reduced data were developed with one of the groups omitted and used for test purposes. Prediction residuals were calculated for each developed model and the process was repeated with another subset of the calibration dataset, until every subset was left out once (Berrueta, Alonso-Salces and Héberger 2007). The final model was that with the lower prediction residual and was used with the external validation dataset in order to perform independent discriminations, evaluating its final performance.

In this study two thirds of the samples (144 spectra of each water stress level, totaling 432 spectra) were used as the calibration and cross-validation dataset, and one third (72 spectra of each water stress level, totalizing 216 spectra) as the external validation dataset. Both datasets contained representative samples of the two experiments and the three water stress conditions. This sampling plan followed the recommended by Kramer (1998) to ensure dataset representativeness. The number of samples used in the calibration process corresponded to more than 10 times the number of variable components in the experiment (water stress level). Each component was considered as an independent source of significant variation in the data.

FDA is a statistical method for determining the most discriminative variables regarding specific category or the category of a sample based on its spectral signature. That is, this method finds linear combinations of variables that separate in the best way predefined categories, deciding in which category a sample should be classified by maximizing the inter-class variance with regard to the total variance (Bourennane et al. 2014). In this study, FDA assessed new variables, called discriminant factors, which were linear combinations of selected PCs resulting from the

PCA analysis, allowing a better separation of the centers of gravity of the considered water stress levels or classes (Devaux et al. 1988). Thus, the distances of each sunflower sample from the centers of gravity associated with the three water stress levels were calculated, assigning the sample with the shortest distance.

PLS-DA also assigns an unknown sample to a predefined class based on its spectral signature and is pointed as suitable for datasets with high degree of inter-correlation between the independent variables (Steidle Neto et al. 2018). According to Ballabio and Consonni (2013), similar to Partial Least Squares Regression for quantification, data reduction is conducted creating latent variables, which are orthogonal with each other, and at the same time trying to describe the predefined classes. In this study, PLS-DA data reduction was also conducted seeking for discriminant factors, which were linear combinations of the original variables, and were calculated in a way to maximize the covariance with the available classes. For this, two matrices (X and Y) were constructed. The X matrix corresponded to the original data, while the Y matrix consisted of three columns associated with the water stress levels and many lines as there were spectra. Each spectrum had the value 1 for the class it belongs to and 0 for the others. During the analysis, a model was developed for each class and the closer a spectrum of a certain column in Y was to 1, the more likely that sample was considered a member of a particular water stress level. This procedure guaranteed that observations were always classified in one of the available water stress levels.

The performances of FDA and PLS-DA models were evaluated by confusion matrices and percentage correct classifications. The confusion matrices represented the numbers of sunflower leaves attributed to each water stress level compared to the reference water contents. The diagonals of the confusion matrices contained the correct classifications, and their numbers were compared to the total number of sunflower samples originally associated with each water stress condition. Confusion matrices were processed considering both calibration with cross-validation and external validation datasets.

The algorithms of the pre-treatments, as well as the calculations required by the multivariate methods, were included in the SCILAB software (Scilab Enterprises, Versailles, France).

RESULTS AND DISCUSSION

Figure 2 presents the average raw spectrum of sunflower leaves after applying centering and normalizing pre-processing, and the same spectrum using detrend, first and second derivative methods.

Centering and normalizing are considered a default first step when multivariate methods are applied for classification purposes, ensuring that all results will be interpretable in terms of variation around the mean (Cozzolino et al. 2011) and neutralizing the influence of undesired noise or drift between analyses (Cynkar et al. 2010). Applying these techniques to the raw sunflower spectra and those pre-treated with detrend, first and second derivatives allowed that the multivariate methods were run based on a single spectral reflectance range (-0.4 to 0.4%). When centering and normalizing were not applied, the same raw spectra results in reflectances ranging from 0 to 30%, while spectral reflectances of sunflowers only treated by detrend vary between -14 and 10% When considering first and second derivatives, spectral reflectances ranged from -0.06 to 0.14% and from -0.003 to 0.004%, respectively.

All the tested pre-treatments enhanced the characteristics of centered and normalized raw spectra, evidencing its peaks and valleys. Compared with detrend, first and second derivatives presented larger effect on baseline, making the peaks and valleys easier to distinguish and increasing the spectra apparent resolution. But first derivative better accentuated the differences of the spectral bands (Figure 2).

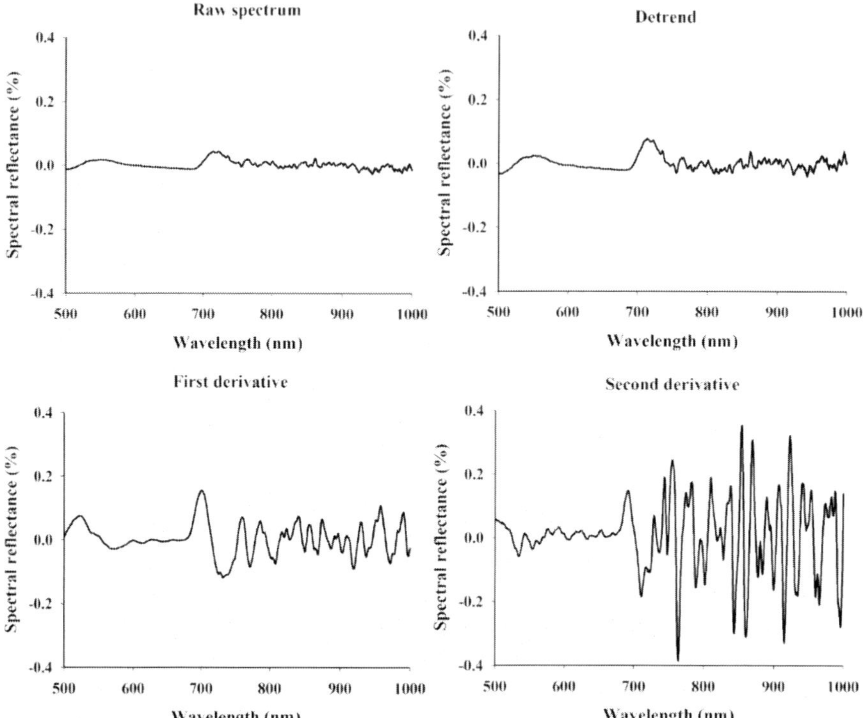

Figure 2. Average spectra of sunflower leaves, after centering and normalizing pre-processing, considering the raw reflectances and those pre-treated with detrend, first and second derivatives.

Models for one to five discriminant factors were investigated, considering that models with fewer discriminant factors tend to be less complex and present a faster processing. Thus, for each multivariate method and spectral pre-treatment, the number of discriminant factors was incremented, and a model was built, until calibration and cross-validation errors did not vary or were greater than the previous analysis. The optimal classifications occurred with two discriminant factors for both supervised methods and all spectral pre-treatments, except for PLS-DA with raw spectra, which required four discriminant factors.

Tables 1 and 2 summarize the results of FDA and PLS-DA models, considering the different spectral pre-treatments and the raw spectra, for calibration with cross-validation in terms of correct classifications and confusion matrices. The overall correct classification of each method and spectral pre-treatment was calculated by averaging the individual classification percentiles of each water stress level. In general, the results of both multivariate methods were very similar. PLS-DA was a little better than FDA when using the raw spectra and applying the second derivative pre-treatment, while FDA performed slightly better for detrend and first derivative procedures.

The highest percentage of correctly classified samples was obtained for leaves without water stress for all spectral pre-treatments and multivariate methods (Tables 1 and 2). Further, none of the leaves without water stress was erroneously classified as under severe stress. Samples under moderate water stress resulted in higher misclassification rates. This reflects the fact that leaves under moderate water stress share similar properties with both other classes considered in this study (without and severe).

Table 1. Confusion matrices resulting from FDA analysis for the calibration with cross-validation considering the different spectral pre-treatments and the raw spectra

Water stress	Without	Moderate	Severe	% correct
Raw spectra – Average accuracy: 77.61%				
Without	135	40	7	93.86
Moderate	9	99	36	68.97
Severe	0	5	101	70.00
Detrend – Average accuracy: 79.27%				
Without	133	37	7	92.11
Moderate	11	102	29	70,69
Severe	0	5	108	75.00
First derivative – Average accuracy: 81.27%				
Without	131	27	0	91.23
Moderate	13	112	36	77.59
Severe	0	5	108	75.00
Second derivative – Average accuracy: 78.94%				
Without	124	22	0	85.96
Moderate	20	109	36	75.86
Severe	0	12	108	75.00

Table 2. Confusion matrices resulting from PLS-DA analysis for the calibration with cross-validation considering the different spectral pre-treatments and the raw spectra

Water stress	Without	Moderate	Severe	% correct
Raw spectra – Average accuracy: 79.56%				
Without	134	37	7	92.98
Moderate	10	102	29	70.69
Severe	0	5	108	75.00
Detrend – Average accuracy: 78.69%				
Without	133	40	7	92.11
Moderate	11	99	29	68.97
Severe	0	5	108	75.00
First derivative – Average accuracy: 80.69%				
Without	141	40	7	98.10
Moderate	11	99	29	68.97
Severe	0	5	108	75.00
Second derivative – Average accuracy: 79.54%				
Without	129	25	0	89.47
Moderate	15	107	36	74.14
Severe	0	12	108	75.00

Discrimination results for external validation using FDA and PLS-DA with raw and pre-treated spectra are presented in Tables 3 and 4, respectively.

When considering the external validation datasets, PLS-DA was a little better than FDA for the raw spectra, as well as for all spectral pre-treatments. As expected, using the models with an independent dataset resulted in slightly lower overall correct classifications, following the same discrimination trends observed during the calibration with cross validation. Quelal-Vásconez et al. (2019) detected cocoa shell in cocoa powder by using NIR spectroscopy, and also noted a similar classification pattern between calibration and external validation datasets, with the last one presenting accuracy 7.5% lower than the first one. Costa, Trugilho and Hein (2018) affirmed that the statistics of the cross-validation model were similar to those of the external validation when classifying eucalypt charcoal quality by NIR spectroscopy, with external validation resulting in accuracies slightly lower than cross-validation.

External validation confirmed the satisfying performance of using both multivariate methods, with raw and pre-treated spectra, for classifying the sunflower plants according to different water stress levels. The percentage of correctly classified samples, as well as the overall accuracies found with the independent dataset, indicated good generalization capacity of the proposed models. The PLS-DA using the first derivative pre-treatment was the most effective for classifying the water stress levels of sunflower leaves, reaching overall correct classification of 80.67% when considering the external validation (Table 4). The lower accuracy was verified when applying the FDA with second derivative, which resulted in an overall correct classification of 69.05% (Table 3).

Table 3. Confusion matrices resulting from FDA analysis for the external validation considering the different spectral pre-treatments and the raw spectra

Water stress	Without	Moderate	Severe	% correct
Raw spectra – Average accuracy: 73.63%				
Without	66	23	7	91.23
Moderate	6	47	19	65.17
Severe	0	2	46	64.50
Detrend – Average accuracy: 78.55%				
Without	66	19	4	91.67
Moderate	6	50	14	68.97
Severe	0	3	54	75.00
First derivative – Average accuracy: 79.85%				
Without	66	17	0	91.75
Moderate	6	53	19	73.79
Severe	0	2	53	74.00
Second derivative – Average accuracy: 69.05%				
Without	62	13	0	86.05
Moderate	10	52	37	72.59
Severe	0	6	35	48.50

Table 4. Confusion matrices resulting from PLS-DA analysis for the external validation considering the different spectral pre-treatments and the raw spectra

Water stress	Without	Moderate	Severe	% correct
Raw spectra – Average accuracy: 76.83%				
Without	66	23	4	91.76
Moderate	6	47	16	65.50
Severe	0	2	53	73.23
Detrend – Average accuracy: 78.08%				
Without	66	21	4	92.29
Moderate	6	48	14	66.96
Severe	0	3	54	75.00
First derivative – Average accuracy: 80.67%				
Without	65	15	0	90.62
Moderate	7	55	18	76.27
Severe	0	2	54	75.13
Second derivative – Average accuracy: 73.96%				
Without	64	12	0	88.30
Moderate	8	53	28	73.08
Severe	0	7	44	60.50

Several studies showed that PLS-DA combines the benefits of other supervised methods with noise reduction and variable selection advantages of traditional PLS method. These characteristics tend to be improved when this method is applied with first derivative spectral pre-treatment, which enhances information on the wavelength bands related to the chemical composition and structure of plants. Grasel and Ferrão (2016) applied the first derivative for a better interpretation of the spectra, when using the PLS-DA method to classify natural tannin extracts by NIR spectroscopy, reaching model sensitivity and specificity of 100%. The effects of using first and second derivatives, standard normal variate, and their combinations were compared by Biancolillo et al. (2018) when developing PLS-DA models to authenticate an Italian hazelnut by NIR spectroscopy, and the most appropriate pre-treatment was the first derivative.

Considering the calibration with cross-validation and the external validation datasets, both PLS-DA and FDA led to satisfactory prediction abilities by using raw or pre-treated sunflower leaf spectra. This suggests that the Vis/NIR spectroscopy coupled with chemometrics is suitable for a fast and non-destructive discrimination of water stress levels in sunflower plants.

CONCLUSION

Vis/NIR spectral reflectance coupled with FDA or PLS-DA multivariate statistical methods applied to centered and normalized raw or pre-treated (detrend, first and second derivative) spectra was shown to be a promising tool for the discrimination of different water stress levels in sunflower plants. Among the methods and pre-treatments tested, PLS-DA using first derivative spectra, performed slight better for classifying the sunflower plants.

This technique can be used into processing lines, enabling large-scale individual analysis and real-time decision making, also contributing for developing irrigation and selection strategies, as well as improving the management of water use efficiency by sunflower plants.

ACKNOWLEDGMENT

The authors are grateful to the Foundation for Research Support of the State of Minas Gerais (FAPEMIG) in Brazil, which provided funding to acquire the spectrometer and accessories (grant numbers: CAG-APQ-01715-13 and CAG-APQ-01495-15).

REFERENCES

Alvares, C. A., Stape, J. L., Sentelhas, P. C., Gonçalves, J. L. M. and Sparovek, G. (2013). Köppen's climate classification map for Brazil. *Meteorologische Zeitschrift*, 22, 711-728.

Ballabio, D. and Consonni, V. (2013). Classification tools in chemistry. Part 1: linear models. PLS-DA. *Analytical Methods*, 5, 3790-3798.

Berrueta, L. A., Alonso-Salces, R. M. and Héberger, K. (2007). Supervised pattern recognition in food analysis. *Journal of Chromatography A*, 1158, 196-214.

Best, N. B., Wang, X., Brittsan, S., Dean, E., Helfers, S. J., Homburg, R. and Hasegawa, P. M. (2016). Sunflower 'Sunspot' is hyposensitive to GA3 and has a missense mutation in the DELLA motif of HaDella1. *Journal of the American Society for Horticultural Science*, 141, 389-394.

Biancolillo, A., De Luca, S., Bassi, S., Roudier, L., Bucci, R., Magrì, A. D. and Marini, F. (2018). Authentication of an Italian PDO hazelnut ("Nocciola Romana") by NIR spectroscopy. *Environmental Science and Pollution Research*, 25, 28780-28786.

Bourennane, H., Couturier, A., Pasquier, C., Chartin, C., Hinschberger, F., Macaire, J. J. and Salvador-Blanes, S. (2014). Comparative performance of classification algorithms for the development of models of spatial distribution of landscape structures. *Geoderma*, 219-220, 136-144.

Buriro, M., Sanjrani, A. S., Chachar, Q. I., Chachar, N. A., Chachar, S. D., Buriro, B. and Mangan, T. (2015). Effect of water stress on growth and yield of sunflower. *Journal of Agricultural Technology*, 11, 1547-1563.

Chaves, M. M., Flexas, J. and Pinheiro, C. (2009). Photosynthesis under drought and salt stress: regulation mechanisms from whole plant to cell. *Annals of Botany*, 103, 551-560.

Costa, L. R., Trugilho, P. F. and Hein, P. R. G. (2018). Evaluation and classification of eucalypt charcoal quality by near infrared spectroscopy. *Biomass and Bioenergy*, 112, 85-92.

Cozzolino, D., Cynkar, W. U., Shah, N. and Smith, P. (2011). Multivariate data analysis applied to spectroscopy: potential application to juice and fruit quality. *Food Research International*, 44, 1888-1896.

Cynkar, W., Dambergs, R., Smith, P. and Cozzolino, D. (2010). Classification of tempranillo ines according to geographic origin: combination of mass spectrometry based electronic nose and chemometrics. *Analytica Chimica Acta*, 660, 227-231.

Devaux, M. F., Bertrand, D., Robert, P. and Qannari, M. (1988). Application of multidimensional analyses to the extraction of discriminant spectral patterns from NIR spectra. *Applied spectroscopy*, 42, 1015-1019.

Diago, M. P., Fernandes, A. M., Millan, B., Tardaguila, J. and Melo-Pinto, P. (2013). Identification of grapevine varieties using leaf spectroscopy and partial least squares. *Computers and Electronics in Agriculture*, 99, 7-13.

Gamon, J. A. and Surfus, J. S. (1999). Assessing leaf pigment content and activity with a reflectometer. *New Phytologist*, 143, 105-117.

Grasel, F. S. and Ferrão, M. F. (2016). A rapid and non-invasive method for the classification of natural tannin extracts by near-infrared spectroscopy and PLS-DA. *Analytical Methods*, 8, 644-649.

Kramer, R. (1998). *Chemometric techniques for quantitative analysis*. New York: CRC Press.

Liu, R., Abdelraouf, E. A. A., Bicego, B., Joshi, V. R. and Garcia, A. G. (2018). Deficit irrigation: a viable option for sustainable confection sunflower (*Helianthus annuus* L.) production in the semi-arid US. *Irrigation Science*, 36, 319-328.

Lopes, D. C. and Steidle Neto, A. J. (2018). Classification and authentication of plants by chemometric analysis of spectral data. In: Lopes, J. and Sousa, C. (Eds). *Vibrational Spectroscopy for Plant Varieties and Cultivars Characterization* (pp. 105-125). Amsterdam: Elsevier.

Martens, H. and Naes, T. (1992). *Multivariate calibration*. New York: John Wiley & Sons.

Monazzah, M., Soleimani, M. J., Enferadi, S. T. and Rabiei, Z. (2018). Effects of oxalic acid and culture filtrate of *Sclerotinia sclerotiorum* on metabolic changes in sunflower evaluated using FT-IR spectroscopy. *Journal of General Plant Pathology*, 84, 2-11.

Moros, J. (2010). Vibrational spectroscopy provides a green tool for multi-component analysis. *Trends in Analytical Chemistry*, 29, 578-591.

Moscetti R., Haff, R. P., Stella, E., Contini, M., Monarca, D., Cecchini, M. and Massantini, R. (2015). Feasibility of NIR spectroscopy to detect olive fruit infested by Bactroceraoleae. *Postharvest Biology and Technology*, 99, 58-62.

Moura, L. O., Lopes, D. C., Steidle Neto, A. J., Ferraz, L. C. L., Carlos, L. A. and Martins, L. M. (2016). Evaluation of techniques for automatic classification of lettuce based on spectral reflectance. *Food Analytical Methods*, 9, 1799-1806.

Patanè, C., Cosentino, S. L. and Anastasi, U. (2017). Sowing time and irrigation scheduling effects on seed yield and fatty acids profile of sunflower in semi-arid climate. *International Journal of Plant Production*, 11, 17-32.

Polin, J. P., Gu, Z., Humburg, D. S. and Dalsted, K. J. (2014). Source of airborne sunflower dust generated during combine harvester operation. *Biosystems Engineering*, 126, 23-29.

Quelal-Vásconez, M. A., Lerma-García, M. J., Pérez-Esteve, É., Arnau-Bonachera, A., Barat, J. M. and Talens, P. (2019). Fast detection of cocoa shell in cocoa powders by near infrared spectroscopy and multivariate analysis. *Food Control*, 99, 68-72.

Sánchez, M. T., Garrido-Varo, A., Guerrero, J. E. and Pérez-Marín, D. (2013). NIRS technology for fast authentication of green asparagus grown under organic and conventional production systems. *Postharvest Biology and Technology*, 85, 116-123.

Saporta, G. (2006). *Probabilités, analyse des données et statistique [Probability, data analysis and statistics]*. Paris: Editions Technip.

Savitzky, A. and Golay, M. J. E. (1964). Smoothing and differentiation of data by simplified least squares procedures. *Analytical Chemistry*, 36, 1627-1639.

Shao, Y., Xie, C., Jiang, L., Shi, J., Zhu, J. and He, Y. (2015). Discrimination of tomatoes bred by spaceflight mutagenesis using visible/near infrared spectroscopy and chemometrics. *Spectrochimica Acta Part A: Molecular and Biomolecular Spectroscopy*, 140, 431-436.

Steidle Neto, A. J., Lopes, D. C. and Borges Júnior, J. C. F. (2017a). Assessment of photosynthetic pigment and water contents in intact sunflower plants from spectral indices. *Agriculture*, 7, 1-9.

Steidle Neto, A. J., Lopes, D. C., Pinto, F. A. C. and Zolnier, S. (2017b). Vis/NIR spectroscopy and chemometrics for non-destructive estimation of water and chlorophyll status in sunflower leaves. *Biosystems Engineering*, 155, 124-133.

Steidle Neto, A. J., Lopes, D. C., Silva, T. G. F., Ferreira, S. O. and Grossi, J. A. S. (2017c). Estimation of leaf water content in sunflower under drought conditions by means of spectral reflectance. *Engineering in Agriculture, Environment and Food*, 10, 104-108.

Steidle Neto, A. J., Lopes, D. C., Toledo, J. V., Zolnier, S. and Silva, T. G. F. (2018). Classification of sugarcane varieties using visible/near infrared spectral reflectance of stalks and multivariate methods. *The Journal of Agricultural Science*, 156, 537-546.

Steidle Neto, A. J., Zolnier, S., Marouelli, W. A., Carrijo, O. A. and Martinez, H. E. P. (2005). Avaliação de um circuito eletrônico para medição da condutividade elétrica de soluções nutritivas [Evaluation of an electronic circuit for measuring the electrical conductivity of nutrient solutions]. *Engenharia Agrícola*, 25, 427-435.

Su, W. H., He, H. J. and Sun, D. W. (2017). Non-destructive and rapid evaluation of staple foods quality by using spectroscopic techniques: a review. *Critical Reviews in Food Science and Nutrition*, 57, 1039-1051.

Symonds, P., Paap, A., Alameh, K., Rowe, J. and Miller, C. (2015). A real-time plant discrimination system utilizing discrete reflectance spectroscopy. *Computers and Electronics in Agriculture*, 117, 57-69.

Ullah, S., Skidmore, A. K., Groen, T. A. and Schlerf, M. (2013). Evaluation of three proposed indices for the retrieval of leaf water content from the mid-wave infrared (2-6 μm) spectra. *Agricultural and Forest Meteorology*, 171-172, 65-71.

Wang, L. and Mizaikoff, B. (2008). Application of multivariate data-analysis techniques to biomedical diagnostics based on mid-infrared spectroscopy. *Analytical and Bioanalytical Chemistry*, 391, 1641-1654.

Westad, F. and Marini, F. (2015). Validation of chemometric models - a tutorial. *Analytica Chimica Acta*, 893, 14-24.

Yuan, L., Huang, Y., Loraamm, R. W., Nie, C., Wang, J. and Zhang, J. (2014). Spectral analysis of winter wheat leaves for detection and differentiation of diseases and insects. *Field Crops Research*, 156, 199-207.

In: Sunflowers
Editor: Érico de Sá Petit Lobão

ISBN: 978-1-53617-195-2

Chapter 5

GENETIC VARIABILITY IN SUNFLOWER AFTER MUTAGENIC TREATMENT OF SEEDS AND IMMATURE EMBRYOS

A. I. Soroka[1] and V. A. Lyakh[2,*]
[1]Institute of Oilseed Crops of the National Academy of Agrarian Sciences, Zaporozhye, Ukraine
[2]Zaporozhye National University, Zaporozhye, Ukraine

ABSTRACT

The spectrum and frequency of morphological and physiological mutations after treatment of mature and immature seeds as well as immature embryos of different age (9-11 and 14-16 days old) with chemical mutagen ethyl methanesulfonate (EMS) were studied. Mature and immature (20 days old) seeds were soaked in the mutagen solution of 0.1 and 0.01% concentration. Immature embryos were treated with 0.02% of EMS. Immature seeds and embryos were later grown in vitro on the modified MS medium. It has been proved that the use of immature embryos for mutagenic treatment, as opposed to mature seeds, resulted in wider genetic variability. That was manifested in increasing the frequency of mutations up to two-fold and broadening the mutation spectrum.

* Corresponding Author's Email: Lyakh@iname.com.

Generation of some specific mutations was greatly influenced by the age of the embryos. Thus, mutations that cause the disruption of male and female generative structures were observed when treating the 14-16 days old embryos only. Agronomic importance of the selected mutant accessions was demonstrated. The emphasis was made on the use of morphological mutants as a source of the marker traits. Genetic control of some mutant traits was revealed.

Keywords: sunflower, mutagenesis, mature and immature seeds, immature embryos, frequency and spectrum of mutations

INTRODUCTION

Among the oilseed crops grown in the Ukraine, sunflower is a priority. The high interest of farmers in this species lies in the universality of the crop itself - it is practically waste-free cultivation with high profitability. One hectare of sunflower field with a seed yield of 2.5 t/ha produces 1200 kg of oil, 800 kg of meal, 500 kg of husk, 1500 kg of heads, 25-30 kg of honey and some other products.

Sunflower yield in France, Argentina or the USA is higher than in the Ukraine. However, despite this, the country ranks second in the world in terms of cultivation of this crop, and among the countries exporting sunflower oil - first.

Growing sunflower on lands close to the Ukraine has a long history. In Russia, it appeared in the second half of the 18th century initially, as in Europe, as an ornamental plant. Later, it was grown in vegetable gardens and seeds consumed as delicacy instead of nuts (Vasiliev 1990). At the beginning of the 19th century, mass popular breeding promoted the emergence of large-seeded accessions used as snack food, and somewhat later, in the thirties, the production of sunflower oil started.

In the Ukraine, sunflower began to be grown as a field crop in the middle of the 19th century. After the October Revolution of 1917, largely due to the creation of a wide network of scientific institutions specializing in sunflower, and because of the breeding of varieties adapted for cultivation in various natural zones, the area under sunflower significantly expanded.

And the use of heterosis breeding has further strengthened the position of this crop in countries with traditional areas of its cultivation.

In recent years, in the Ukraine there has been a steady growth of its acreage in all growing areas. Nowadays fields under this oilseed crop occupy more than 20% of the total sown area of the country, or more than 6 million hectares, which is two times more than it was in 2008. And in the sunflower belt, which includes Dnepropetrovsk, Kirovograd, Zaporozhye, Nykolaev, and a number of other regions of the Ukraine, acreage proportion amounted to 30% and above.

Large planted areas under sunflower, leading to overcropping, its promotion to the regions with different agro-ecological conditions pose new challenges to breeders, the solution of which begins with the search for the original germplasm. That is why the problem of increasing genetic variability is so acute, since the existing variability is almost exhausted and does not meet the modern demands of the breeder.

History of Induced Mutagenesis in Sunflower

For many years induced mutagenesis has been a reliable method for extending genetic variability in numerous crops. One of the first researches on the induced mutagenesis in sunflower was the work of Soldatov (Soldatov 1968). He obtained a mutant accession of sunflower with an oleic acid content of about 70%, which gave rise to the high-oleic variety Pervenets. This mutant formed the basis of many modern varieties and hybrids with a high content of oleic acid.

With the help of chemical mutagenesis, a series of mutants of sunflower with modified architectonics and a shortened vegetation period was developed. Some plants, for example, due to the changes in the length of leaf petioles, had the shape of spruce or column, which allowed them to be used in thickened crops. On the basis of one of the mutant lines, they managed to create accessions in which the ratio of vegetative and generative organs was 56 to 44 compared to the original 70 to 30 (Kalaijyan 1991).

Using N-nitroso-N-methylurea (NMU), nitroso-ethylurea, dimethyl sulfate, ethyleneimine and 1,4-bis-diazoacetylbutane, 45 types of morphological and other mutations were induced (Surovikin and Soldatov 1978). As a result of the action of nitroso-ethylurea and dimethyl sulfonate on seeds of the variety VNIIMK 8931, a group of mutants was obtained which differed from the control in plant height and length of the vegetation period. The height of the mutants varied from 20 to 250 cm at 150 cm in the control, while the dwarfs possessed such an important breeding trait as a strong non-lodging upright stem. Among the mutants with a modified vegetation period, there were fast ripening, early ripening, and ultra early-ripening samples (Kalaijyan and Soldatov 1991).

A number of researchers have reported obtaining plastid mutants of sunflower. Some of the plastid mutants were characterized by a complex of economically important traits. A sunflower hybrid was obtained, one of the parental components of which is the drought-resistant plastome mutant created by chemical mutagenesis (Beletsky and Razoriteleva 1984). Studying the effect of NMU on sunflower, some patterns of mutation induction were revealed. Exposure to the mutagen in small doses (0.01 and 0.015%; 3 hours) resulted mainly in the appearance of extranuclear chlorophyll mutations. Increasing the concentration to 0.02% leads to the induction of not only plastid, but also nuclear chlorophyll mutations. The mutagen at a concentration of 0.03% causes major structural rearrangements of nuclear genetic material, which are recorded as chromosomal aberrations (Usatov et al. 2011).

It has been established that effect of the mutagenic action for a number of chemical compounds can be modified, in particular by hyperbaric oxygenation. It was shown that, when applied to the plastid genetic apparatus of sunflower, it increased the efficiency of nitroso-methylurea (Usatov et al. 2005).

There are some pieces of information about the induction of mutations in the tissue culture of sunflower. In this way, lines resistant to fomopsis and with improved economic performance, in particular, increased oil content, were identified (Encheva et al. 2003). Other authors have isolated and studied a genetic mutation that affects the content of photosynthetic pigments (Fambrini et al. 2003). Attempts were also made to obtain Alternaria-resistant sunflower plants using in vitro mutagenesis, where the callus induced from the explants was subjected to EMS treatment (Binodh et al. 2008). J. Encheva et al. (2008) treated the isolated immature zygotic embryos with ultrasound before plating them on a nutrition medium in vitro. The genetic changes obtained included 15 morphological and biochemical agronomic traits.

As for the action of physical mutagens on sunflower, in this regard studies are known by Indian scientists who have obtained a fairly wide range of mutations resulting from the treatment of sunflower seeds with gamma rays (Jambhulkar and Joshua 1999). Among the visible morphological changes, the authors observed mutations of leaves, ray and tubular flowers, seeds, as well as a large number of chlorophyll changes.

S. Gvozdenovic and co-authors determined the role of genetic differences in the response of sunflower to induced mutagenesis in order to determine the optimal treatment conditions. Seeds of 15 sunflower genotypes were treated with gamma rays, fast neutrons and EMS at various doses. Based on the damaging effect of a mutagen on the height of seedlings they concluded that the 50% damage indices varied from 120 to 329 Gy for gamma rays, from 9 to 21 Gy for fast neutrons and from 0.69 to 1.55% for EMS concentration (Gvozdenovich et al. 2008).

Indian researchers obtained a sunflower mutant with 125 leaves, compared with 30-35 for the parental line and a dwarf mutant about 11 cm high with a height of 180 cm for the parent. Both mutations were controlled by a single recessive gene (Jambhulkar and Shitre 2008).

An attempt was made to induce variability in sunflower for resistance to *Alternaria helianthi*. Treating seeds with gamma-rays (150 and 165 Gy) and EMS (0.015 mol dm^{-3}) M.F. de Oliveira et al. (2004) recovered M_2, M_3 and M_4 plants with no or less than 5% diseased leaf area.

A number of biochemical mutants in sunflower with altered oil quality was obtained. The most valuable of them were characterized by a high content of oleic or linoleic acid (more than 80-90%), a high content of palmitic, palmitoleic or stearic acid (more than 25%), a low total amount of saturated fatty acids, and a high content of different tocopherols. The mutagens used to treat seeds included both physical and chemical agents - gamma- and X-rays, sodium azide, dimethyl sulfate, ethyl methanesulfonate. New traits in all cases were controlled by a small number of genes, which facilitates their involvement in the breeding process (Fernandez-Martinez et al. 2008; Skorić et al. 2008).

With the help of induced mutagenesis, mutants of sunflower were obtained with an increase in 3-5 times, compared to the original inbred line, absorption of cadmium, zinc and lead from the soil. It is assumed that the obtained accessions can be successfully used for phytoextraction of toxic metals (Nehnevajova et al. 2008).

Kumar and Ratnam (2010) studied the joint action of gamma-rays and sodium azide, as well as their separate effects on seed germination, seedling survival, pollen fertility and seed set in sunflower. They found that in combined treatments, mutagenic effectiveness gradually increased with an increased dose.

Cvejić et al. (2011) irradiated seeds of eight sunflower inbred lines with gamma rays (γ) and fast neutrons and treated in ethyl methanesulfonate (EMS) solution. As a result they isolated in M_2 and M_3 7 mutants - early flowering, short and high plant stature, higher oil content and branching.

Kumar et al. (2013) using EMS mutagenesis found that about 1.1% of the plants in the M_2 sunflower population showed phenotypic variations. They also calculated, based on the TILLING of two genes, that the overall mutation rate to one mutation occurred every 480 kb.

Zambelli et al. (2015) note the importance of induced mutagenesis to improve sunflower, particularly emphasizing its significance for the progress of the sunflower oil market and in the creation of herbicide-resistant genotypes.

However, despite the apparent large number of inherited changes resulting from induced mutagenesis, the question of expanding the gene pool for this important crop is still quite acute. The limited source material affects both the genetic uniformity of sunflower varieties and hybrids transferred to production fields, and the insufficient knowledge of the private genetics of this species.

Results and Discussion

Influence of EMS Treatment of Mature and Immature Seeds on Genetic Variability in Sunflower

Characteristics of M_1 Plants

Both mature and immature seeds were treated with a mutagen. Mature seeds of three lines from Zaporozhye Institute of Oilseed Crops ZL9, ZL102 and ZL169 were soaked in an aqueous solution of ethyl methanesulfonate at the concentrations of 0.01%, 0.1%, 0.5% during 6 and 12 hours. After treatment with the mutagen, the seeds were washed under running water for 30 minutes and immediately planted in the field. In the control the seeds were soaked in distilled water.

As immature seeds, seeds of 20 days old age of the lines ZL-102 and ZL-169 were used. They were soaked in an aqueous solution of the mutagen at the concentrations of 0.01% and 0.1% during 6 and 12 hours. In the control seeds of 20 days old age were soaked in water at the same exposure level. After soaking, the seeds were washed for 30 minutes, sterilized in sodium hypochlorite solution for 15 minutes, and rinsed with sterile water (Lyakh et al. 2005). The seeds were then placed on a modified MS nutrient medium with a reduced number of micro- and macroelements (Murashige and Skoog 1962). After 7-10 days, seedlings with a branched root system were planted in a pre-disinfected soil in half-liter plastic vials and subsequently grown in a phytotron with a 16/8 h day/night photoperiod.

Two types of changes were identified on plants of the M_1 generation (Table 1). One of them was the deformation of the leaves (modified leaf shape, lack of part of the leaf, the difference in growth between the right and left halves of the lamina, sprouting veins, necrosis). With an increase in both the mutagen concentration and the time of exposure, an increase in the number of plants with deformed leaves occurs.

Table 1. Effect of sunflower immature seeds treatment with ethyl methanesulfonate on M_1 plants

EMS exposure, h	EMS concentration, %	Plants with chlorophyll sections, %	Plants with chlorophyll sections and sprouting veins, %	Plants with deformed leaves, %
ZL169				
6	Control	0	0	14.2
	0.01	3.1	0	37.5*
	0.1	27.4***	8**	29.0
12	Control	0	0	10.3
	0.01	2.4	0	43.9**
	0.1	45.4***	19.6***	96.9***
ZL102				
6	Control	0	0	0
	0.01	0	0	4.1
	0.1	0	0	2.5
12	Control	0	0	0
	0.01	0	0	4.1
	0.1	24.1***	0	12.0**

Note. *, **, *** differences from the control are significant at $p < 0.05$, 0.01, and 0.001, respectively.

Especially well this pattern can be traced with the line ZL169. If in the control there was no more than 14% of leaves with deformations and they were located mainly in the lower belt, then with 0.1% EMS treatment and at exposure of 6 hours one third of plants had deformed leaves, and with an increase in exposure to 12 hours almost all the plants of this line demonstrated such leaves. At the same time, they were located not only in the lower but in the upper belt as well.

Among the deformed leaves such morphoses as sprouting veins were noted. Sprouting veins in the ZL102 line were formed after the action of mutagen on immature seeds and were absent in the control. They were detected only after mutagen treatment at the maximum concentration and exposure. In the ZL169 line under the action of mutagen, compared with the control, there was considerably more plants with sprouting veins.

Chlorophyll chimeras were another type of changes. They were formed only under the action of mutagen on immature seeds and were completely absent in the controls of both lines. When treating with a 0.1% solution of ethyl methanesulfonate and an exposure of 12 hours, both genotypes revealed the maximum number of chimeric plants. Thus, the line ZL169 produced about half of such plants. With a decrease in exposure time from 12 to 6 hours and the same EMS concentration, the number of chimeric plants in the ZL169 line lowered 1.6 times, and in the ZL102 line they were not found at all (Table 1).

After treatment with EMS at the concentration of 0.01%, both at exposure of 6 and 12 hours, one chimeric plant was found for the line ZL169, and such plants were not found at all for the line ZL102 (Table 1) .

The main types of chimeras (chlorophyll-deficient sections on leaves), we found in M_1:

by location on the plant

- on one leaf only
- on two or more leaves
- several sectors on one leaf

by location on a leaf blade

- at the base of the leaf blade
- along the leaf blade margin
- at the top of the leaf blade
- narrow strip along the central vein
- the section occupies half of the leaf
- the section is located only on the upper side of the leaf
- the section is located only on the lower side of the leaf
- the section is located on both sides of the leaf

by color

- white
- yellow
- yellow-green
- light green

Characteristics of chimeric plants derived in M_1 generation, listed in the Table 2, illustrates well the differences in the effect of EMS at various concentrations on immature seeds.

Table 2. Proportion of sunflower plants differing in chimeras type in M_1 generation, %

Exposure, h	EMS mutagen, %	Number of chimeric sections per leaf		Location of chimeric sections on the plant		Color of chimeric sections	
		one	several	on one leaf	on two and more leaves	yellow green or light green	white or yellow
ZL169							
6	Control	0	0	0	0	0	0
	0.01	100	0	100	0	100	0
	0.1	73.6	26.4	58.8	41.2	62.5	37.5
12	Control	0	0	0	0	0	0
	0.01	100	0	100	0	100	0
	0.1	80	20	70	30	72.9	27.1
ZL102							
6	Control	0	0	0	0	0	0
	0.01	0	0	0	0	0	0
	0.1	0	0	0	0	0	0
12	Control	0	0	0	0	0	0
	0.01	0	0	0	0	0	0
	0.1	78.5	21.5	78.5	21.5	70.5	29.5

For example, the line ZL169 at a lower mutagen concentration (0.01%), regardless of the exposure time, had only one section with chlorophyll deficiency per each chimeric plant and it was colored light green. With an increase in EMS concentration to 0.1%, both at exposure of 6 and 12 hours, from 30 to 40% of the plants had chlorophyll-deficient sections, which were located on two or more leaves. At the same time, about a quarter of the plants demonstrated more than two chimeric sections per single leaf.

With a high concentration of EMS, the plants were in proportion in the color of chimeras as follows: 60-70% of plants of the line ZL169 showed chlorophyll-deficient sectors with a color ranging from light green to yellow-green, and the rest of plants had a yellow or white color. As for the line ZL102, the chlorophyll-deficient sections of various colors were found only at the maximum concentration and exposure of the mutagen treatment. At the same time, the ratio of plants with different coloring of chimeras in both lines was similar.

When comparing the data we obtained with data from other researchers, both similarities (leaf deformation, chlorophyll-deficient areas on the leaf blade) and differences in the effect of mutagens on M_1 plants were found. The difference includes such a morphose as sprouting veins of a leaf blade that is not described in the literature.

As a result of the treatment of sunflower immature seeds with EMS, the number of observed changes in M_1 plants was several times higher than after treatment of mature seeds with any chemical mutagen or radiation carried out by other authors. Thus, for example, the number of chlorophyll chimeras was 15 times higher than in the treatment of mature seeds with gamma radiation (Jambhulkar and Joshua 1999). The high frequency of morphoses, revealed in the M_1, suggested the derivation of a significant number of genetically modified plants in future generations.

Mutation Spectrum in M_2

The intensity of the mutation process is characterized not only by the frequency, but also by the spectrum of identified mutations. In M_1, only those mutations that are dominant occur. Most mutations are recessive and cannot manifest themselves until they become homozygous. In our studies, no plants with mutant traits were found in the M_1. All mutants were identified in the M_2.

The total number of visible mutation types, which were identified after treatment of mature and immature seeds, amounted to 19 for all three lines, of which 17 were induced by the mutagenic treatment of mature seeds, 5 – when the immature seeds were used (Table 3).

Table 3. Spectrum of mutations in M_2 under the action of EMS on mature and immature sunflower seeds

No.*	Type of mutation
	I. Chlorophyll mutations
1	White seedlings (*albina*)
2	Yellow seedlings and plants, the color of which subsequently turns green (*virescent*)
3	Light green seedlings and plants (*viridis*)
4	Light yellow upper leaves (*xantha*)
5	White or yellow spots on individual leaf segments (chimera)
6	White-green upper leaves and randomly scattered white sectors on all leaves (whitish)
	II. Mutations of shape and color of ray flowers
7	Narrow and short ray flowers
8	Light-yellow color of ray flowers
9	Lemon color of ray flowers
	III. Mutations of stem shape and structure
10	Low stem
11	Dwarfs
12	Fasciated stem
13	Malformed growth point
	IV. Mutations of leaf shape and structure
14	Fan-nerved venation
15	Sickle curved leaf blade
16	Egg-shaped leaves
17	Corrugated leaves
18	Fringed margin of the leaf blade
	V. Mutations in physiological characteristics of plant growth and development
19	Fast maturity

Note. *The numbers for mutant types indicated in this table correspond to the numbers in the following tables.

Six types of chlorophyll mutations were isolated in the M_2 generation, *albina*, chimera, *virescent, viridis, xantha* and “whitish”. The first five types are often found in mutation studies on sunflower and other crops. Besides those mutation types, “whitish” plants were found. Such mutants have white-green upper leaves and white spots on the leaves which remain for the whole vegetation period. This mutation essentially reduces the height of plant and decreases plant productivity. Chlorophyll mutation of *xantha* type was also uncommon. Except white-green upper leaves surrounding the capitulum the bottom of the head was yellow-green until ripening.

One mutation of ray floret shape and two ray floret color mutations were isolated in the present study. The first mutation was characterized by narrow and short ray florets. After EMS treatment, ray florets of lemon-yellow and light-yellow color were found in contrast to the source lines which possessed ray florets of yellow color.

Four mutations were assigned to the group of stem mutations - short-growing, dwarfness, fasciation and with distorted growing point. The fasciated mutant, with a flattened stem, shortened internodes and a large number of small leaves, was a typical example of fasciation mutation. Low-growing mutants included plants whose reduced height was due to a smaller number of internodes. Dwarf plant had very short internodes. The stem of that mutant was covered with a large number of whitish hairs. The mutants with distorted growing point did not develop normally and did not set seeds.

In the present study, 5 distinct types of morphological mutations involving shape of leaf lamina and leaf venation were isolated. Sickle-shaped, egg-shaped, fringed and corrugated leaf mutations and the mutation of fan leaf venation were found. The sickle-shaped leaf mutant, because of arched central vein, has one half of the leaf lamina shorter than another.

Figure 1 shows the ratio of mutation groups for three sunflower lines when treating mature seeds with EMS. The percentage indicates the number of mutant types in this group in relation to the total number of mutation types detected in the line.

All studied lines have the largest group of mutations with impaired chlorophyll synthesis. The values for the other groups varied depending on the genotype.

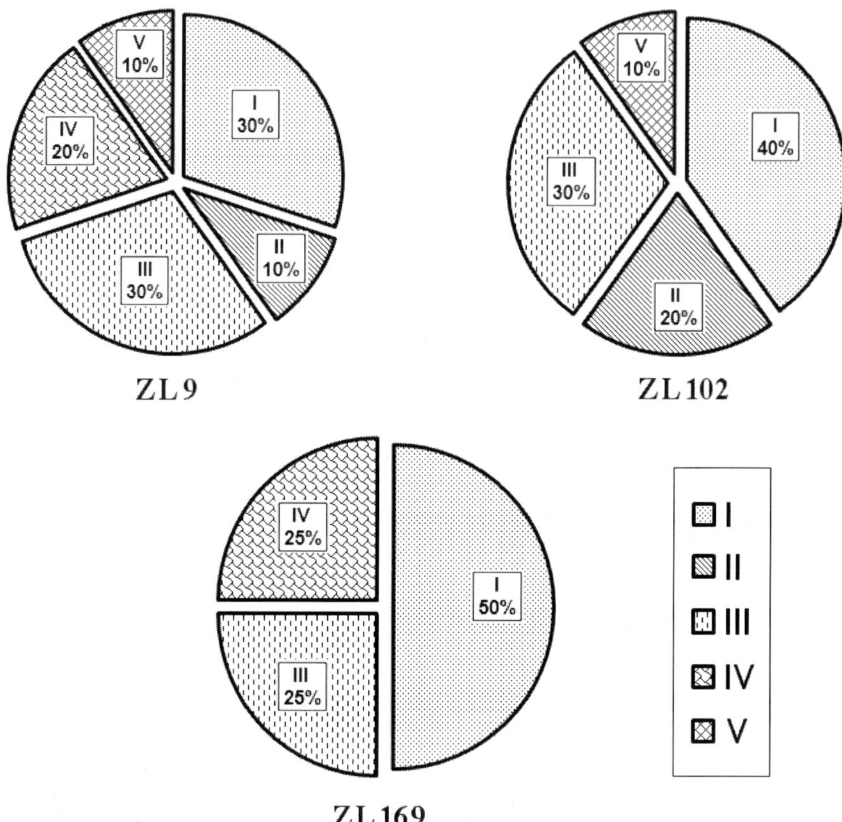

Figure 1. Ratio of mutation groups induced by EMS treatment of mature seeds in different sunflower lines: I. Chlorophyll mutations, II. Mutations of ray florets shape and color, III. Mutations of stem shape and structure, IV. Mutations of leaf shape and structure, V. Mutations in physiological characteristics of plant growth and development.

It should be noted that complex mutations have also been identified that affect several traits. Thus, in the mutant with fan leaf venation, the shape of ray flowers changed, in the mutant with chlorophyll disruption of whitish type, the leaf blade changed from normal to sickle curved, and the plant dwarfism mutation was combined with the fringed margin of the leaf blade. It is characteristic that when immature seeds were treated with the mutagen, complex mutations constituted a significant part of the mutation spectrum.

Mutation Frequency in M_2

The Frequency of Visible Mutations in M_2 under the Action of Ethyl Methanesulfonate on Mature Seeds

The intensity of the mutation process is characterized not only by the spectrum, but also by the frequency of mutations. We estimated the mutation frequency in three sunflower lines (ZL169, ZL9, and ZL102), which is summarized in the general Table 4.

Chlorophyll mutations were isolated with a rather high frequency, although different, depending on the concentration of the mutagen and exposure time. The lethal *albina* (1) mutation was detected in all three lines. It was found at both exposures and practically only at the maximum (0.5%) concentration of ethyl methanesulfonate. The *virescent* (2) type was isolated in the same way as the *albina* mutation at the highest mutagen concentrations and only at 12 hour exposure. Mutations of *viridis* (3) and *xantha* (4) types were rare. A mutation of the chimera type (5) was detected in all lines, with the highest frequency in the ZL169 line. The whitish (6) type was encountered only in the line ZL169 and was found at all the treatments. With a small frequency it was also isolated in the control treatment of this line.

Mutations in the shape and color of ray flowers were found in two of the three lines. They were detected mainly at the maximum (0.5%) mutagen concentration. Narrow and short ray florets were isolated at the 6 hour exposure of the mutagen, and light yellow color of ray florets was found at both exposures. Lemon color of ray florets was detected with a softer mutagen treatment (EMS concentration - 0.01%, exposure time - 6 hours).

Low stem plants (10) were detected in all three lines with different treatments. The fasciation of the stem (12) and plants with a deformed growth point (13) were also found at different concentrations of EMS and different exposures.

Mutations of leaf shape and structure were genotype specific. Fan-nerved venation (14) was isolated only in the line ZL9 with the maximum concentration of mutagen and 6 hour exposure. Plants with ovoid leaves (16) were detected at the same line at two minimum concentrations of EMS and 6-hour exposure.

Sickle curved leaf blade (15) was detected with a high frequency (from 2.0 to 5.7%) only in the line ZL169 with all treatments. Also, a mutation of corrugated leaves was typical for this line, found at 12 hour exposure and 0.01% EMS.

Table 4. The frequency of mutations in M_2 after EMS treatment of mature seeds of three sunflower lines, %

Type of mutation	Exposure time of mutagen treatment							
	6 hours				12 hours			
	0.00%	0.01%	0.10%	0.50%	0.00%	0.01%	0.10%	0.50%
1	0.00	0.00	0.36	1.43	0.00	0.37	0.00	2.60
2	0.00	0.00	0.00	0.00	0.00	0.00	0.00	0.74
3	0.00	0.38	0.00	0.00	0.00	0.00	0.00	0.00
4	0.00	0.00	0.36	0.00	0.00	0.00	0.00	0.00
5	0.00	1.52	1.45	1.07	0.00	1.86	1.06	1.86
6	0.34	0.76	0.72	2.14	0.00	1.86	0.71	0.37
7	0.00	0.00	0.00	0.36	0.00	0.00	0.00	0.00
8	0.00	0.00	0.00	0.36	0.00	0.00	0.00	0.37
9	0.00	0.38	0.00	0.00	0.00	0.00	0.00	0.00
10	0.00	0.38	0.36	0.36	0.00	1.12	0.35	0.37
11	0.00	0.00	0.00	0.00	0.00	0.00	0.00	0.00
12	0.00	0.38	0.00	0.36	0.00	0.37	0.71	0.00
13	0.00	0.38	0.36	0.36	0.00	0.37	0.71	0.37
14	0.00	0.00	0.00	0.36	0.00	0.00	0.00	0.00
15	0.00	0.76	0.72	2.14	0.00	1.86	0.71	0.37
16	0.00	0.38	0.36	0.00	0.00	0.00	0.00	0.00
17	0.00	0.00	0.00	0.00	0.00	0.37	0.00	0.00
18	0.00	0.00	0.00	0.00	0.00	0.00	0.00	0.00
19	0.00	0.38	0.00	0.00	0.00	0.37	0.00	0.37
Number of families	293	263	276	280	297	269	282	269
Total frequency	0.34	5.70	4.71	8.93	0.00	8.55	4.26	7.43

In general, both the spectrum and frequency of mutations were rather genotype-specific. The highest frequency of mutations was characteristic of the line ZL169 with a maximum of 17.6%. For all treatments, this line demonstrated a significant number of families with inherited changes.

The higher mutability of one line compared to another indicates the decisive role of the genotype in inducing ethyl methanesulfonate mutations in sunflower. This is confirmed by the results of other authors' researches to identify the effect of different mutagens, belonging to the class of alkylating compounds, on sunflower.

High frequencies of visible chlorophyll mutations were also expected. This is explained by the specificity of the action of ethyl methanesulfonate, which in many crops causes a large number of mutations with a change in the color of vegetative organs of the plant (Agrawal et al. 2005; Muthusamy et al. 2005).

Frequency of Visible Mutations in M_2 under the Action of Ethyl Methanesulfonate on Immature Sunflower Seeds

Studies with immature seeds were carried out on two lines - ZL102 and ZL169. After the treatment with ethyl methanesulfonate of immature seeds of the ZL102 line, mutations were not detected either at exposure of 6 or 12 hours, and they were found in the ZL169 line only at a longer 12 hours exposure (Table 5).

A higher (0.1%) concentration of the mutagen led to the emergence of five types of mutations, and the concentration of an order of magnitude lower (0.01%) - only three. For both treatments, the total frequency of inherited changes ranged from 14.0% to 15.2%.

Chlorophyll mutations accounted for about half of the inherited changes. Of the six types of chlorophyll mutations found after the treatment of mature seeds, in the case of immature seeds two types were identified – *xantha* (4) and whitish (6). The *xantha* (4) mutation with a frequency of just over 2% was detected at both mutagen concentrations. When treating mature seeds, it was rarer and found only in one of the M_2 families with an EMS concentration of 0.1%. Chlorophyllic mutant of whitish type (6), as well as *xantha* (4), was detected in both treatments of immature seeds but with a greater frequency (4.0 - 6.5%). Together with the whitish mutation (6) and with the same frequency the mutant type of the sickle curved leaf blade was detected (15).

At the mutagen concentration of 0.1% and maximum exposure time, an original complex mutation was found, combining mutant types of dwarf (11) and fringed leaf margin (18). When treating mature seeds, this type of mutation was not observed.

Thus, treatment of immature seeds with ethyl methanesulfonate, followed by *in vitro* cultivation, did not lead to the emergence of inherited changes in M_2 in the line ZL102, but induced morphological mutations with a total frequency of up to 15.2% in the line ZL169. In this case, mutations were detected only with a greater (12 hour) exposure. A significantly lower sensitivity to the mutagen of the line ZL102, as compared to ZL169 was also found during the treatment of mature seeds with ethyl methanesulfonate. It should be noted that under the action of the mutagen on immature seeds, rare original mutations were isolated, which were detected with a low frequency or were not detected at all when processing mature seeds.

Table 5. Frequency of mutations in M_2 when treating immature seeds of the line ZL169 with ethyl methanesulfonate, %

Type of mutation	Exposure time					
	6 hours			12 hours		
	0.0%	0.01%	0.1%	0.0%	0.01%	0.1%
4	0	0	0	0	2.17	2.0
6	0	0	0	0	6.52	4.0
11	0	0	0	0	0	2.0
15	0	0	0	0	6.52	4.0
18	0	0	0	0	0	2.0
Number of families	17	24	44	25	46	50
Total frequency	0	0	0	0	15.2 ± 5.29	14.0 ± 4.90

EMS Treatment of Immature Embryos of Different Age

As already noted, to induce mutational variability mature seeds are most often subjected to mutagenic treatments. In Section 3 we demonstrated that EMS treatment of immature sunflower seeds has led to the appearance of inherited changes that were not detected when mature seeds were subjected to mutagen action. Immature embryos can also be used for mutagenic treatment (Checheneva and Larchenko 1997). In this case, a new spectrum of mutations is expected due to a different gene expression.

Characteristics of Plants in M_1 Generation

For the study the immature embryos of two sunflower lines of Zaporozhye breeding (ZL809 and ZL95) were used as the material. After flowering, the plants were self-pollinated, and after 9-11 and 14-16 days, seeds with immature embryos were isolated from the heads. Then seeds were subjected to the action of 0.02% aqueous solution of EMS for 16 hours, dehulled, and derived embryos were planted under sterile conditions on a modified MS nutrient medium with a reduced content of macronutrients and an increased content of vitamins. In the control, seeds with immature embryos were treated with distilled water instead of EMS. Subsequently, the embryos were cultured in Petri dishes at a 16-hour photoperiod and room temperature until germination. Green seedlings were planted first in plastic cups, and after the emergence of two or three pairs of true leaves - in the field (Soroka and Lyakh 2009).

To assess the effectiveness of the influence of the mutagenic factor on embryos, the survival rate of the obtained plants was evaluated (Table 6). As can be seen, after treatment with ethyl methanesulfonate, significant embryo death was observed, as compared with the control, both at the age of 9-11 and 14-16 days. However, the mutagen most adversely affected the embryos of the youngest age. The main cause of the embryo death was a lack of the embryonic root, apparently due to damage of the root meristem cells.

Plants that survived embryo treatment with the mutagen, started to bloom with a significant delay, 15 to 17 days on average, compared with the control plants. The vegetation period was prolonged by 4-12 days as well. The mutagen, however most strongly affected plant height. EMS also influenced head diameter and the number of seeds per head, differently for both lines. However, in general M_1 plants in all treatments set a sufficient amount of seeds to carry out further studies on determining the frequency and spectrum of mutations.

Table 6. Survival rate of sunflower plants after treatment of immature embryos of different ages with ethyl methanesulfonate

Age of embryos, days	Treatment	Embryos treated	Plants in field	Productive Plants	Survival of productive plants, %
ZL809					
14-16	Control	75	59	51	68.0 ± 5.39
	EMS	238	169	115	48.3 ± 3.24
9-11	Control	73	70	8	11.0 ± 3.66
	EMS	224	0	0	0
ZL95					
14-16	Control	67	36	34	50.7 ± 6.11
	EMS	243	77	65	26.7 ± 2.84
9-11	Control	123	61	41	33.3 ± 4.25
	EMS	165	34	32	19.4 ± 3.08

Changes of the quantitative traits in M_1 plants as a result of the embryo treatment with ethyl methanesulfonate indicated a significant effect of the mutagen on this object and made it possible to hope for obtaining a high mutation rate and a wide range of inherited changes in subsequent generations. In general, it should be noted that embryos of a younger age require a less rigid background treatment. The same applies to the treatment of immature sunflower embryos with gamma rays, which in our preliminary experiments in doses of 100-400 Gy were lethal for embryos of both ages.

Mutation Spectrum in M_2 and M_3 Induced by the Influence of EMS on Immature Sunflower Embryos

Each M_2 family, consisting of about 30 plants, was the offspring of a single M_1 plant. The M_3 family was the offspring of a single plant selected from the M_2 family. To search for mutations in the M_3 generation, 4-5 plants with leaf morphoses were selected in M_2. In M_3 only those changes that were not found in the earlier (M_2) generation were taken into account.

No plants with mutant traits were identified in M_1. All mutants were found in the subsequent generations - M_2 and M_3. Treatment with ethyl methanesulfonate of immature embryos of 9-10 and 14-15 days old resulted in a wide spectrum of morphological and physiological mutations in M_2 and M_3. Their brief description is given in Table 7.

All inherited changes were combined into seven groups: chlorophyll deficiency, cotyledon leaves, true leaves, stem, inflorescence, seeds, and physiological traits. The chlorophyll mutations of *viridis* (1) and *xantha* (2) types are common to many crops. They were also found earlier when we treated mature and immature sunflower seeds with the mutagen. In addition to these chlorophyll mutations, mutants with bright yellow spots on the leaves, - *xantha* of necrotic type (3), which turned into necrotic sectors towards the end of the vegetation period, were isolated in one of the variants after mutagen treatment of immature embryos.

Mutations of cotyledonary and true leaves were represented by 7 types of mutations. In a tube-leaf mutant, the first pair of true leaves was fused and formed a tube-like structure. As the plant grew, new appearing leaves have torn the tube but its presence on the plant could be recognized even by the end of the growing season. Mutation of fan venation affected a number of traits. Those plants were characterized by the absence of a clearly defined central vein on the leaves, fringed leaf blade and the erect arrangement of the leaves.

Table 7. Spectrum of mutations in M_2 and M_3 under the action of ethyl methanesulfonate on immature sunflower embryos

No.	Type of mutation	Characteristics
1	2	3
I. Chlorophyll mutations		
1	Viridis	Light green seedlings and plant
2	Xantha	Yellow-green seedlings and plant
3	Xantha of necrotic type	Yellow-green seedlings and plant, bright yellow-green spots on the leaves, turning into necrotic sections by the end of the growing season
II. Mutations of cotyledon leaves		
4	Deformed cotyledons	Twisted, often fused or dissected, cotyledon leaves
III. Mutations of true leaves		
5	Fringed leaf	Very wavy leaf margin
6	Tube-shaped leaf	The first pair of true leaves fused into a tube
7	Deformed leaf	Twisted, often fused or dissected, leaf blades
8	Large leaf	Leaf blade of big size
9	Fan-nerved venation	Fan-nerved venation, fringed leaf, smaller angle between stem and leaf petiole
10	Erectoid leaf	Smaller angle between stem and leaf petiole
IV. Mutations of stem		
11	Low-growing plants	Reduced plant height
12	Tall plants	Increased plant height
13	Low-growing plants with strong habit	Low-growing plants with strong stem and large leaves
14	Tall plants with strong habit	Tall plants with strong stem and large leaves
15	Strong habit	Plants with strong stem and large leaves
16	Tilted stem	The top of the plant with a head is bent almost to the ground by the end of the growing season
17	Tobacco- shaped plant	Plant with short internodes, rounded leaves, wide and rounded cotyledons, reduced number of short ray flowers
18	Fasciated stem	Flattened fused stem
19	Branching	1-3 side shoots in the basal part of the stem
V. Mutations of inflorescence		
20	Few bracts	Reduced number of bracts
21	Many bracts	Increased number of bracts
22	Malformed bracts	Significant overgrowth of bracts

No.	Type of mutation	Characteristics
23	Capitulum fasciation	Deformed capitulum, fusion of smaller heads
24	Head inclination angle	Head is directed vertically up
25	Partial capitulum sterility	Underdevelopment of generative organs in the center of the head
26	Full sterility	Underdevelopment of male and female generative structures, lack of ray flowers
27	Few ray flowers	Reduced number of ray flowers
28	Many ray flowers	Increased number of ray flowers
29	Short ray flowers	Smaller length of ray flowers
30	Corrugated ray flowers	Curved, wavy-edged ray flowers
VI. Mutations of seed		
31	Seed color	Red-brown color of the seed coat, white at the beginning of ripening
VII. Physiological mutations		
32	Early flowering	Shortened by 2-4 days germination-flowering period
33	Late flowering	Elongated by 2-4 days germination-flowering period

Among plants with stem mutations, one should distinguish mutants with a modified habit (short, tall, strong) and a tilted stem. Tobacco-like mutant had obvious differences from normal plants starting from the seedling stage.

Mutations of the inflorescence included changes in the number of ray flowers and bracts, the shape of ray flowers, the inclination angle and shape of the head, as well as mutations of sterility.

After mutagenic treatment of immature embryos one type of seed mutations was detected. The mutants had a red-brown color of the seed coat, while in the control seeds were black.

The range of mutations in both M_2 and M_3 was quite wide. Despite the fact that the lines under study differed significantly in the mutation spectrum, it depended on the age of embryos as well. For example, treatment of only embryos of 14-15 days old from the line ZL95 resulted in the emergence of mutations of complete sterility. Those mutants were characterized by a complete disruption in the development of male and female generative structures, therefore we maintained them through heterozygotes. In general, the spectrum of visible changes in the M_3

generation reflected the spectrum of mutations in the M_2 generation, although some traits were unique to the M_3, since they were not detected in the previous generation.

Mutation Frequency in M_2 and M_3 That Was Induced by EMS Treatment of Sunflower Immature Embryos

Data on the frequency of visible inherited changes in M_2 confirmed the significant effect of the mutagen on immature embryos, which was expressed in the appearance of a significant number of mutations in both generations. As can be seen from Figure 2, a higher mutation rate was characteristic of the line ZL95. That index in this line exceeded the one of the line ZL809 by 3-5 times. Obviously, this is due to significant differences in the genetic origin of the two accessions studied.

No differences in the total mutation frequency between the variants with different ages of the embryos were found. However, between those variants there were differences in the frequency of mutations of individual types. Thus, after mutagenic treatment of the embryos of 14-16 days old from the line ZL95 the most common mutations were the habit mutations. Their frequency amounted almost to 15%. When EMS was applied to younger (9-11 days old) embryos of this line, mutations of the bracts, as well as branching mutations, were detected with a high rate.

Usually in M_3 generation the mutation rate is lower compared to M_2. In our case, when immature embryos were treated with the mutagen, the frequency of inherited changes in M_3 did not differ significantly from that in the earlier generation. This pattern was characteristic of both lines.

As already noted, in the M_2 generation, the spectrum of mutations differed in treatments with different age of immature embryos. However, the greatest differences in the mutation spectrum between those treatments were observed when analyzing the M_3 generation. Thus, after the treatment of the youngest embryos of the line ZL95 such rare mutations as fan leaf venation, tobacco-like plant, as well as xantha were only found in this variant. When embryos of this line of 14-16 days old were treated a high rate of sterility mutations with a complete disruption in the development of male and female generative structures was registered.

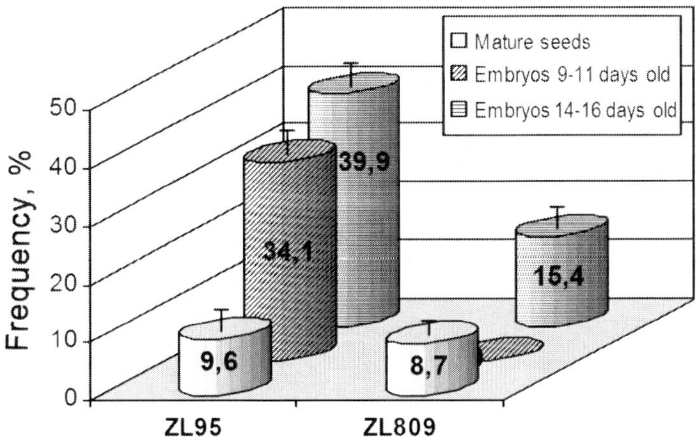

Figure 2. Mutation frequency in M_2 after EMS treatment of sunflower immature embryos and mature seeds.

ECONOMIC VALUE OF OBTAINED SUNFLOWER MUTANTS

Mutants with Marker Traits

In the course of our studies a number of chlorophyll-deficient mutants with a change in the color of vegetative parts of the plant were identified. Those chlorophyll mutations differed significantly both in their external manifestation and in the degree of plant depression.

Table 8 demonstrates the data on the effect of a number of such mutations on the growth and development of sunflower plants. The table shows that mutations *viridis* and *xantha*, in contrast to *virescent* and whitish, did not significantly affect the studied parameters. This suggests the possibility of using chlorophyll mutations *viridis* (3, seedlings and adult plants have a light green leaf color) and *xantha* (4, the top four to six leaves at the stage of head have a yellow-green color that persists until the end of the growing season) as marker traits in sunflower seed production.

Table 8. Characteristics of some economically important traits in plants of the control and mutant sunflower lines

Line	Plant height, m	Number of leaves	Head diameter, cm	1000 seed weight, g	Vegetation period, days
Control, ZL102	1.3	25.8	20.3	47.3	105
Viridis	1.2	27.2	23.6	46.1	103
Virescent	0.8**	25.0	20.8	42.8	107
Light-yellow color of ray florets	1.2	23.4	23.2	47.0	105
Lemon color of ray florets	1.3	26.4	26.0	47.5	105
Control, ZL169	0.8	17.6	23.6	45.4	93
Xantha	0.7	22.8	25.2	44.6	94
Whitish	0.5*	20.0	15.79	43.5	90

*,** Differences from control are significant when $p < 0.05$ and 0.01, respectively.

Inherited color changes in ray flowers of sunflower were identified and studied from the beginning of the twentieth century. Already then it was found out that, along with the usual yellow color of ray flowers, there was variability from red to white color (Fick 1976). In our studies mutations of ray flowers have also been isolated. The line ZL102 gave rise to two mutant lines, the main difference of which was the color of ray flowers - light yellow and lemon, in contrast to the yellow in the original line.

Mutants with light yellow and lemon color of ray flowers did not differ from the control by all economically valuable traits. And only during flowering, due to a clear difference in color, it could be possible to notice difference between them and the source line. That allows to recommend those mutations, similar to the chlorophyll mutations *viridis* and *xantha*, for use as marker traits.

Agronomically-Important Mutants

Sunflower is characterized by a significant variation in the stem height. Although the degree of variability in the height largely depends on external conditions, the main factor that determines plant height is the genotypic characteristic of the accession.

In our studies a treatment with the mutagen led to a change in this morphological trait. Three types of mutations were identified - plants with low stem, plants with tall stem and dwarfs. The most interesting among them were the dwarfs. A dwarf plant with a height of about 50 cm was detected in the line ZL169 after EMS treatment of immature seeds. In this case the dwarfism was due to a strong shortening of internodes, and the stem as a consequence of heavy pubescence had an almost white color.

The presence of these attributes in the mutant prompted studies of the anatomical and morphological structure of the leaves, which are considered by some authors as diagnostic indicators of the adaptability to arid environments (Khristova 1999). For this purpose we examined the structure of the leaf cover tissue in the dwarf and in the source line ZL169 (Table 9).

As can be seen, the mutant line in the structure of the leaf integumentary tissue differs from the original line in a large number of trichomes, a large number and smaller size of the main cells and the guard cells of the stomata. These differences in the structure of the leaf epidermis give reason to believe that the dwarf mutant line is more adapted to tolerating dry conditions than the source line ZL169.

In addition to the dwarf mutant isolated from the ZL169 line as a result of treating immature seeds with the mutagen, plants with a reduced height were also detected after mutagenic treatment of immature embryos (Table 10).

Table 9. Characteristics of abaxial side epidermis of sunflower leaf in the line ZL169 and its dwarf mutant

Trait	ZL169 line	Dwarf mutant
Leaf cell length, μm	54.3	39.3*
Stomata guard cell length, μm	31.0	22.3**
Number of cells per mm2	1223.0	2074.4***
Number of trichomes on leaves, pcs/cm2	66.8	106.3*

*, **, *** - differences are significant at $p < 0.05$, 0.01 and 0.001, respectively.

Table 10. Characteristics of mutants with altered plant height

Line	Treatment	Plant height, cm	Number of leaves
ZL169	Control	80.0 ± 8.00	17.6 ± 1.82
ZL169 dwarf mutant	0.1% EMS, 12h, immature seeds	50.0 ± 6.00	18.6 ± 1.52
ZL95	Control	111.8 ± 1.29	20.1 ± 0.49
ZL95 “tobacco-like” mutant	0.02% EMS, 16h, 9-11 days old immature embryos	95.8 ± 3.20	26.3 ± 0.94
ZL809	Control	102.6 ± 3.06	19.9 ± 0.54
ZL809 low height mutant	0.02% EMS, 16h, 14-16 days old immature embryos	81.3 ± 2.24	24.9 ± 0.87

In contrast to the dwarf line, the undersized mutants derived from the lines ZL95 and ZL809 were characterized by a significantly larger number of leaves, which provides the formation of a larger amount of photosynthesis products. Simultaneous change in the plant height and the number of leaves significantly altered the habit of such plant, thereby expanding the utilization possibilities of the crop.

Use of Morphological Mutants to Determine Possible Transitional Links in the Process of Evolution of Angiosperms

The leaf of angiosperm is characterized by extreme evolutionary plasticity and polymorphism within the limits of a single family and even of a genus. The shape and structure of the leaf are modified as a result of evolutionary changes in the original species. In sunflower leaf shape is also quite diverse. Thus in the collection of the All-Russian Institute of Plant Industry named after N. I. Vavilov there are 23 variations of the leaf shape.

In our studies we identified 5 types of mutations in the shape and structure of the sunflower leaf. Among them, as a result of mutagen treatment of mature seeds of the line ZL9, a mutation of the fan-nerved venation was identified. In our opinion, this mutation could be of interest as an example of a link in a sequential chain of leaf evolution during phylogenesis. The most striking distinguishing feature of this mutant in comparison with the control plants is the type of leaf venation. In this regard, we examined how the venation changed during the evolution of the angiosperms and how the mutant line can fit into a series of successive changes.

In the process of evolution of leaf venation, both the ratio of lateral to central veins, and the arrangement of the veins in a leaf blade were changed. Regarding the first trait, the leaves of the initial line ZL9 were characterized by pinnate venation. In the mutant, a transitional type from pinnate to a fan-like venation was observed. Its central vein is not so pronounced as in the control line, and the most lateral veins of the first order does not start from the central vein, but from the petiole, with the lateral veins branching dichotomously and forming anastomoses. The angle of deviation of the lateral veins from the implicitly pronounced central vein is also changed. In the mutant it is more acute and averages 30-35°, while in the control it is 60-70°. A sharper deviation angle of veins is characteristic of more primitive types of pinnate venation (Takhtajan 1964).

In accordance with the classification of Takhtajan (Takhtajan 1964), leaves of the line ZL9 have a type of venation close to the reticulated (dictiodromic) one, when the lateral veins, not reaching the edge of the leaf blade, repeatedly split and their numerous branches join together, forming a net of separate loops. In the mutant with fan-nerved venation there are also features of reticulated venation. However, there are also signs of craspedodromous venation, when some lateral veins reach to the margin of the leaf and, quite often, even protrude over the edge in the shape of teeth or bristles.

In general, the line ZL9 possesses the most modern type of venation, while the mutant accession has the most primitive features. The obtained mutation can be easily placed into a hypothetical range of intermediate types of venation and leaf blade shapes.

Thus, the identified mutant can be a good example of the transition from one type of leaf venation to another and can be a transitional link from ancestral to modern shapes and indicate possible phylogenetic relationships.

INHERITANCE OF NEW MUTANT TRAITS

Changes in the color of vegetative parts of plants comprised a large group of mutations that we obtained when sunflower mature and immature seeds and embryos were treated with ethyl methanesulfonate. It is known that chlorophyll mutations in sunflower occur with a rather high frequency under the action of nitroso-methylurea, caffeine (Soldatov 1968), gamma rays (Jambhulkar and Joshua 1999) and some other mutagens. Mutations of this type can arise as a result of disruptions in both the genome and plastome (Beletsky et al. 1990). Employees of Vavilov Research Institute of Plant Industry (St. Petersburg, Russia) carried out the work to study the patterns of genetic inheritance of leaf color trait in sunflower samples with yellow-green, light green, green and dark green color. It was found that the inheritance of leaf color in sunflower is rather complex and diverse (Gavrilova and Anisimova 2003).

We studied the inheritance of traits in mutants with different manifestations of chlorophyll deficiency - *viridis*, *virescent*, *xantha*, whitish, and *xantha* of necrotic type. To clarify whether the chlorophyll mutations identified by us are nuclear or cytoplasmic in origin, and to establish the nature of the inheritance of these mutant traits, we performed reciprocal crosses of those mutants with their source lines. The first generation hybrids in all variants of crossing, except for those with the participation of the *xantha* of necrotic type mutant, had a green color of leaves, and in the F_2 showed 3 : 1 (green : chlorophyll deficiency) ratio. This indicated the genes which determine chlorophyll changes like *viridis*, *virescent*, whitish, and *xantha* are nuclear, and the mutant traits are recessive and inherited monogenically in crosses with the source lines. The F_1 hybrids with the participation of the *xantha* of necrotic type mutant differed in leaf color, which indicated the plastome localization of this mutation.

In the course of studies on induced mutagenesis we found two types of modified leaf venation which were called as fan-nerved venation. These mutants differed in appearance of the leaf central rib. The F_1 plants from crosses between those mutants and between the mutants and genotypes with reticulate leaf venation had the usual reticulate leaf venation. The F_2

segregated into four phenotypic classes for leaf venation in the ratio of 9 reticulate (double dominant phenotype):3 fan-venation type 1:3 fan-venation type 2:1 combined fan-venation (double recessive homozygote). Those results demonstrated that two non-allelic recessive genes (*vf1* and *vf2*) with complementary interaction are involved in the genetic control of fan-nerved leaf venation, and reticulate venation was determined by a combination of at least two dominant alleles of these genes (Soroka and Lyakh 2015).

In the line ZL809 a sample was selected with a reduced number of bracts. We crossed this mutant (25 bracts on average) with another accession (79 bracts on average) to obtain F_1 hybrids. The F_1 hybrids had an average value between parents, and in the F_2 population we found variation from 20 to 84 bracts. Assuming that the differences between the parental lines were due to two pairs of genes, the F_2 plant population was divided into 5 classes. In that population the observed ratio of classes turned out to be close to the theoretically expected ratio of 1: 4: 6: 4: 1. This gave reason to talk about the two-loci control of a such quantitative trait as the number of bracts.

Another trait, the inheritance of which we studied, was the trait of seed shape, evaluated as a length to width (L/W) ratio. The selected mutant with rounded shape of the seeds with close to 1.0 L/W ratio was crossed to elongated seed line with L/W ratio equal to 1.6-1.7. Seeds of F_1 hybrids were intermediate as compared to their parents, however, still elongated. The chi-square test indicated that the F_2 segregation corresponded to the theoretically expected both 15:1 and 63:1 ratios. This indicated the control of the seed shape by two or even three unidirectionally acting genes (Soroka et al. 2017).

After mutagenic treatment of sunflower immature seeds and immature embryos the plants with obtuse shape of leaf tip and strongly sinuate (fringed) leaf margin were found. In crossings between these mutants, all F_1 plants had wild phenotype (acute leaf tip, flat leaf margin). In F_2 generation 1/16 of plants with obtuse leaves and 1/4 of fringed plants were found. It was concluded that two genetic loci with duplicate genes interaction without cumulative effect are responsible for the shape of leaf tip. The trait of fringed leaf was recessive in relation to the normal serrate leaf (absence of fringe) and inherited monogenically. Both traits of leaf are inherited independently (Soroka and Lyakh 2017).

Conclusion

This chapter presents the results on the induction of genetic variability in sunflower when mature and immature seeds and immature embryos were treated with the chemical mutagen ethyl methanesulfonate. Taking into account the fact that different sets of genes are expressed during plant ontogenesis, we assumed that treating in an ontogenetic sense "younger" seed material than mature seeds, one can expect another genetic diversity than after standard seed treatment. The data obtained confirmed our assumption. First of all, this deals with the differences not so much in the mutation frequency but in the mutation spectrum. Even when immature embryos of different ages were mutagenized, the spectra of mutational variability turned out to be in many respects specific for each case. Due to the treatment of immature seeds and immature embryos with ethyl methanesulfonate, a number of economically valuable genotypes were found, such as, for example, the drought-resistant mutant in the ZL169 line or the tobacco-like plant in the ZL95 line. Similar mutants were not detectable in the case of mature seeds. In general, the conducted research demonstrates a high level of genetic variability generated by induced mutagenesis. Therefore, mutation breeding obviously will remain one of the mainstays of modern sunflower breeding.

References

Agrawal, P. K., Sidu, G. S., and Gossal, S. S. (2005). "Induction of bacterial blight resistance in elite Indian rice (*Oryza sativa* L.) cultivars using gamma irradiation and ethyl methane sulfonate." *Mutation Breeding Newsletter & Reviews* 1:16-17.

Beletsky, Y. D. and Razoriteleva, E. K. (1984). "Hybrid form of sunflower, obtained on the basis of a drought-resistant plastome mutant." In *Chemical mutagenesis in increasing the productivity of agricultural plants*, edited by IA Rapoport, 152-155. Moscow: Nauka.

Binodh, A. K., Vindhyavarman, P., Manivannan, N. and Gnanam, R. (2008). "In vitro mutagenesis for Alternaria resistance in sunflower (*Heliantus annuus* L.)." *Paper presented at the FAO/IAEA International Symposium on Induced Mutations in Plants*, Vienna, Austria, August 12-15.

Checheneva, T. N. and Larchenko, E. A. (1997). "Influence of chemical mutagens and gamma irradiation on in vitro culture of inbred corn lines." *Cytology and Genetics* 3(3):65-71.

Cvejić, S., Jocić, S., Prodanović, S., Terzić, S., Miladinović, D. and Balalić, I. (2011). "Creating new genetic variability in sunflower using unduced mutations." *Helia* 34(55):47-54.

Encheva, J., Christov, M. and Shindrova, P. (2008). "Developing mutant sunflower line (*Helianthus annuus* L.) by combined used of classical method with unduced mutagenesis and embryo culture method." *Bulg. J. Agric. Sci.* 14(4): 397-404.

Encheva, J., Kohler, H. and Friedt, W. (2003). "Field evalution of somaclonal variation in sunflower (*Helianthus annuus* L.) and its application for crop improvement." *Euphytica* 130(2):167-175.

Fambrini, M., Bertini, D., Salvini, M. and Pugliesi, C. (2003). "Characterization of a pigment-dieficient chlorina mutant of sunflower (*Helianthus annuus* L.) induced by in vitro tissue culture." *J. Genet. and Breed.* 57(1):81-88.

Fernandez-Martinez, J. M., Perez-Vich, B. and Velasco, L. (2008). "Mutation breeding for oil quality improvement in sunflower." *Paper presented at the FAO/IAEA International Symposium on Induced Mutations in Plants*, Vienna, Austria, August 12-15.

Fick, G. N. (1976). "Genetics of floral colour and morphology in sunflower." *J. Heredity* 67:227-230.

Gavrilova, V. A. and Anisimova, I. N. (2003). *Genetics of cultivated plants. Sunflower.* St. Petersburg: Vavilov Research Institute of Plant Industry.

Gvozdenovic, S., Bado, S., Afza, R., Jocic, S. and Mba, C. (2008). "Intervarietal differences in response of sunflower (*Heliantus annuus* L.) to different mutagenic treatments." *Paper presented at the FAO/IAEA International Symposium on Induced Mutations in Plants*, Vienna, Austria, August 12-15.

Jambhulkar, S. J. and Joshua, D. C. (1999). "Induction of plant injury, chimera, chlorophyll and morphological mutations in sunflower using gamma rays." *Helia*, 22(31):63-74.

Jambhulkar, S. J. and Shitre, A. S. (2008). "Development and utilisation of genetic variability through induced mutagenesis in sunflowers (*Heliantus annuus* L.)." *Paper presented at the FAO/IAEA International Symposium on Induced Mutations in Plants*, Vienna, Austria, August 12-15.

Kalajjan, A. A. (1991). "Induced variability in the height of sunflower plants." *Scientific and Technical Bulletin of All-Russian Scientific-Research Institute of Oilseeds* 1(112):3-12.

Kalajjan, K. I. and Soldatov, K. I. (1991). "Induced variability in the length of the growing season in sunflower." *Scientific and Technical Bulletin of All-Russian Scientific-Research Institute of Oilseeds* 1(112):13-18.

Khristova, T. E. (1999). "Structure of corn hybrid leaf epidermis in modeling of drought." *Ukr. Bot. J.* 56(5):531-535.

Kumar, A. P. K., Boualem, A., Bhattacharya, A., Parikh, S., Desai, N., Zambelli, A., Leon, A., Chatterjee, M. and Bendahmane, A. (2013). "SMART – Sunflower mutant population and reverse genetic tool for crop improvement." *BMC Plant Biology* 13:3-8.

Kumar, P. R. R. and Ratnam, S. V. (2010). "Mutagenic effectiveness and efficiency in varieties of sunflower (*Helianthus annuus* L.) by separate and combined treatment with gamma-rays and sodium azide." *African Journal of Biotechnology* 9(39): 6517-6521.

Lyakh, V., Soroka, A. and Vasin, V. (2005). "Influence of mature and immature sunflower seed treatment with ethylmethanesulphonate on mutation spectrum and frequency" *Helia* 28(43):87-98.

Muthusamy, A., Vasanth, K. and Jayabalan, N. (2005). "Induced high yielding mutants in cotton (*Gossypium hirsuthum* L.)." *Mutation Breeding Newsletter & Reviews* 1:6-8.

Murashige, T. and Skoog, F. (1962). "A revised medium for rapid growth and bioassays with tobacco tissue cultures." *Physiol. Plant.* 15:473-497.

Nehnevajova, E., Herzig, R., Schwitzguebel, J. P. and Schmulling, T. (2008). "Sunflower mutants with improved growth and metal accumulation traits show a potential for soil decontamination." *Paper presented at the FAO/IAEA International Symposium on Induced Mutations in Plants*, Vienna, Austria, August 12-15.

Oliveira, M. F., Neto, A. T., Leite, R. M. V. B. C., Castiglioni, V. B. R. and Arias, C. A. A. (2004). "Mutation breeding in sunflower for resistance to Alternaria leaf spot." *Helia* 27(41):41-50.

Skorić, D., Jocić, S., Sakac, Z. and Lecić, N. (2008). Genetic possibilities for altering sunflower oil quality to obtain novel oils. *Can. J. Physiol Pharmacol.* 86(4):215-21. doi: 10.1139/Y08-008.

Soldatov, K. I. (1968). "Effect of chemical mutagens on oilseeds." In: *Use of chemical mutagenesis in plant breeding*, 42-44. Moscow: Nauka.

Soroka, A. and Lyakh, V. (2009). "Genetic variability in sunflower after mutagen treatment of immature embryos of different age." *Helia* 28(43):87-98.

Soroka, A. and Lyakh, V. (2015). "Inheritance of two types of modified leaf venation in sunflower (*Helianthus annuus* L.)." *Indian J. of Genetics and Plant Breeding* 75(1):75-78.

Soroka, A. and Lyakh, V. (2017). "Inheritance of Leaf Tip Shape and Fringed Leaf Margin in Sunflower." *Helia* 40(66):21-28.

Soroka, A., Totsky, I. and Lyakh, V. (2017). "Inheritance of rounded seed shape in sunflower." *Helia* 40(67):189-196.

Surovikin, V. N. and Soldatov, K. I. (1978). "Study of the use of chemical mutagenesis in sunflower breeding." In *Chemical mutagenesis and hybridization*, 44-51. Moscow: Nauka.

Takhtajan, A. L. (1964). *Basics of angiosperm evolutionary morphology.* Moscow, Nauka.

Usatov, A. V., Fedorenko, G. M., Ustenko, A. A., Tikhonov, M. A., Mashkina, E. V., Azarin, K. V., Markin, N. V., Gorbachenko, O. F., Kolokolova, N. S. and Denisenko, Y. V. (2011). *Nonchromosomal mutations of sunflower.* Rostov n/D: Publisher SFU.

Usatov, A. V., Mashkina, E. V. and Guskov, E. P. (2005). "The effect of oxidative stress on mutagenesis in sunflower (*Helianthus annuus* L.) induced by nitrosomethylurea" *Genetics* 41(1):63-70.

Vasiliev, D. S. (1990). *Sunflower.* Moscow: Agropromizdat.

Zambelli, A., León, A. and Garcés, R. (2015). "Mutagenesis in Sunflower." In *Sunflower. Chemistry, Production, Processing and Utilization*, edited by E. Martínez Force, Dunford, N. T. and Salas, J. J. 7:27-52. AOCS Press.

In: Sunflowers
Editor: Érico de Sá Petit Lobão
ISBN: 978-1-53617-195-2

Chapter 6

DEVELOPMENT OF ENVIRONMENTALLY FRIENDLY COMPOSITE MATERIALS BASED ON SUNFLOWER OIL AND ALFA FIBERS

***Sihem Kadem*[1], *Ratiba Irinislimane*[1,2] *and Naima Belhaneche-Bensemra*[1,*]**
[1]Laboratoire des Sciences et Techniques de l'Environnement, Ecole Nationale Polytechnique, El-Harrach, Alger, Algeria
[2]Université M'Hamed Bougara, Faculté de Sciences, Siège (Ex-INIL) Boumerdès, Algeria

ABSTRACT

This chapter discusses the use of sunflower oil as an epoxy resin in biocomposite material. Sunflower oil is characterized by a great proportion of unsaturated fatty acids (oleic, linoleic and linolenic) which confers a high polymerization capacity. For this, sunflower oil was chemically modified by epoxidation and acrylation and then copolymerized to obtain epoxy resin. This new class of bioderived epoxy resins shows a good mechanical and thermal properties and higher resistance to moisture absorption.

---

[*] Corresponding Author's Email: naima.belhaneche@g.enp.edu.dz.

Epoxy resin was used as a matrix and alfa fiber as reinforcement to prepare biocomposites materials which were characterized in terms of mechanical and thermal properties.

A biodegradation study was carried out for the synthesized biocomposites by measuring the CO_2 level obtained by means of a laboratory respirometry test and by the measurement of the mass loss in a solid medium (burial in the soil) during one year.

The results showed that biodegradation of the biocomposites occurred.

Keywords: sunflower oil, epoxidation, acrylation, alfa fibers, biocomposites, biodegradation, respirometry

1. Introduction

In recent years, environmentally friendly composite materials or biocomposites materials prepared from vegetable oils are becoming increasingly important to replace polymeric materials prepared from petroleum-based source in many structural applications. Because of their biodegradability, safety and price, these vegetable oils have received much attention. Linseed, sunflower, castor, soybean, palm, tall and rapeseed are vegetable oils with special fatty acid distribution for each one. These oils can be converted into epoxidized resins by different epoxidation methods and used for synthesis of oil-modified polymers (Seniha Güner et al. 2006; Malacea and Dixneuf 2010; Carbonell-Verdu et al. 2015). Sunflower oil *(Helianthus annuus*) is one of the four most important vegetable oils worldwide (palm oil, soybean oil, rapeseed oil and sunflower oil). Like all other plant oils, the triglyceride molecule is the main component of sunflower oil. It is composed of a glycerol center and three fatty acid chains joined together. The double bonds on the fatty acid chains were used as reactive sites to prepare new composite matrix through various polymerization techniques: cationic, thermal, and free radical polymerizations (Wool et al. 2000;Pfister and Larock 2010). In this chapter the sunflower oil was chemically modified by epoxidation and acrylation and then copolymerized to obtain epoxy resin.

The biocomposite material was prapared by adding alfa fiber also known as *Stipatenacissima* as reinforcement to the synthesized sunflower oil resin. The prepared materials were characterized in terms of tensile and thermal properties. After that a biodegradation study was carried out for the synthesized biocomposites by measuring the CO_2 level obtained by means of a laboratory respirometry test during six months and by the measurement of the mass loss in a solid medium (burial in the soil) during one year.

2. Elaboration of the Biocomposite

The biocomposites were prepared with a resin from sunflower oil and renforced with alfa fiber.

2.1. Resin Preparation from Sunflower Oil

The sunflower (*Helianthus annuus*) culture was first domesticated by the Indians who used it as food, medicine and body paint in ceremonies. Today, sunflower oil is among the most important vegetable oils on the world market. About 90% of all sunflower oil production is used for food, and only 10% for biodiesel production and for industrial purposes. Sunflower is grown on 23 million hectares, with an annual production rate of about 30 million tons of grain (Sánchez-Muniz and Cuesta 2003;Kaya et al. 2012). The largest producers of sunflower are Russia (22%), the European Union (21%), Ukraine (19%) and Argentina (11%). Now, sunflower is grown also in countries where it has not been before, especially in Asia and Africa. This trend in sunflower production is the result of high quality sunflower oil compared to other major oilseed crops. As an industrial plant, sunflower is mainly grown for the purpose of use in the food industry, the production of biodiesel, various lubricants and in the cosmetics industry (Sánchez-Muniz and Cuesta 2003; Vollmann and Rajcan 2009).

2.1.1. Chemical Composition of Sunflower Oil

Sunflower oil comes from sunflower seeds; its chemical composition mainly comprises unsaturated fatty acids that are liquid at room temperature. It is a triglyceride composed of three branches rather than being a single straight chain compound. Each of the three branches is a fatty acid, mostly a fatty acid with 18 carbon atoms. It is rich in acids of the linoleic form (C 18: 2) about 68% and oleic acid (C 18: 1) (21%). In addition, sunflower oil also contains traces of other saturated fatty acids, such as, palmitic acid (C 16: 0), stearic acid (C 18: 0), myristic (C 14: 0) and capric (C 10: 0). All these fats represent about 10% of the total fatty acid content in the oil (Benaniba et al. 2007; Mechakra et al. 2015), this fatty acid content may change depending on the culture zone, which means that the climatic factors (temperature, light sunlight ...) can significantly change the quality of the oil (Faruk et al. 2012). The chemical structure of sunflower oil (Figure 1) is highly unsaturated or polyunsaturated, which shows that this oil has a high polymerization capacity (crosslinking) (Sánchez-Muniz and Cuesta 2003; Olson 2009; Malacea and Dixneuf 2010).

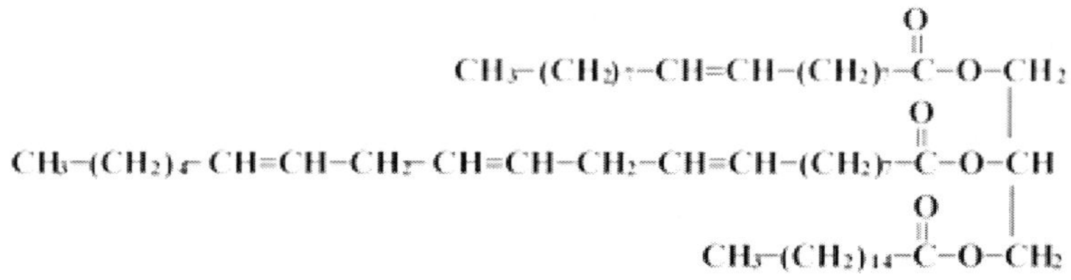

$$CH_3-(CH_2)_7-CH=CH-(CH_2)_7-\overset{O}{\overset{\|}{C}}-O-CH_2$$
$$CH_3-(CH_2)_4-CH=CH-CH_2-CH=CH-CH_2-CH=CH-(CH_2)_7-\overset{O}{\overset{\|}{C}}-O-CH$$
$$CH_3-(CH_2)_{14}-\overset{O}{\overset{\|}{C}}-O-CH_2$$

Figure 1. Structure of sunflower oil (Crivello and Narayan 1992).

2.1.2. Chemical Modification of Sunflower Oil

The triglyceride-based polymers from vegetable oils (sunflower oil, soybean oil, linseed oil, castor oil, etc.) are synthesized using different types of polymerizations (cationic polymerization, chain or ring opening polymerization) to obtain different types of polymers (resins, rubbers, thermosets) and this after several chemical modifications of the structure of these vegetable oils (Kaplan 1998; La Scala and Woo 2005; Irinislimane. and Belhaneche-Bensemra 2012).

2.1.2.1. Epoxidation

The epoxidation of vegetable oils is a grafting of the oxirane ring in the carbon-carbon double bonds (C = C) in fatty acids by conventional epoxidation using formic acid or acetic acid in the presence of the peroxide of hydrogen (H_2O_2), it is one of the most important functionalization processes. Epoxidized vegetable oils, such as epoxidized soybean oil, epoxidized linseed oil, epoxidized canola oil and epoxidized castor oil, have been successfully used to prepare thermosetting polymers and composite materials. The mechanism of epoxidation is the reaction of an alkene with an organic peracid, the carbon-carbon double bond (C = C) is broken giving rise to a triangular cyclic ether (Scheme 2) called epoxide or oxirane, the final oxirane number in this reactions can reach 6.5-7.1% (Kalnin'sh 2010; Malacea and Dixneuf 2010; Irinislimane. and Belhaneche-Bensemra 2012).

The epoxidation of sunflower oil was carried out by reacting this oil with hydrogen peroxide (H_2O_2) in the presence of formic acid (HCOOH) as a catalyst at 50°C. The reactor, consisting of a tri-necked flask equipped with a condenser and a thermometer, is subjected to magnetic stirring and continuous temperature control using a thermocouple. The synthesized called epoxidized sunflower oil (ESO) has an oxirane number between 5.8 and 6.5 (Benaniba et al. 2007; Benaniba et al. 2008; Kadem et al. 2018).

2.1.2.2. Acrylation

The acrylate monomers were prepared by reacting the epoxy groups of the epoxidized oils and epoxidized triglycerides with acrylic acid.

Once modified by acrylic acid, the epoxidized oil has more functional groups following grafting of the acrylate groups on triglycerides (alcohols, esters and C=C double bonds) as shown in Figure 2.

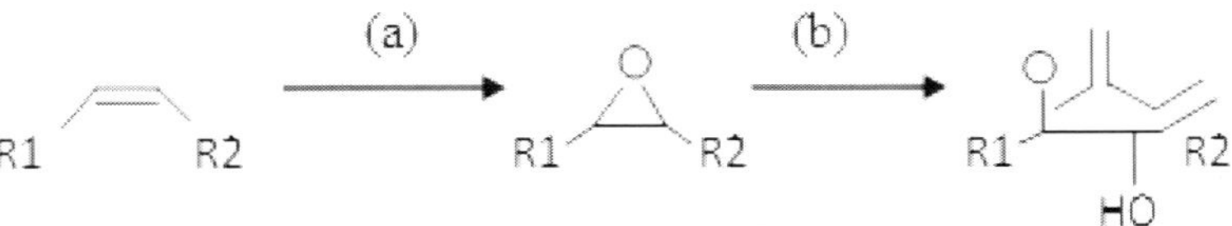

*R1 and R2: Hydrocarbon groups with open chain.

Figure 2. Diagram of epoxidation reaction (a) and acrylation reaction (b) in triglycerides.

The acrylation reaction of the epoxidized sunflower oil (ESO) was carried out in a two-necked flask equipped with a condenser by reacting the epoxidized sunflower oil with acrylic acid in the presence of triethylamine ($N(CH_2CH_3)_3$)as a catalyst. The following acrylatedepoxidized sunflower oil has an oxirane index of less than 2 (Irinislimane. and Belhaneche-Bensemra 2012; Kadem et al. 2018).

2.1.2.3. Polymerization and Copolymerization

Polymerization is a chemical reaction between two or more molecules to give a larger molecule. This is the process by which plastics and resins form. The polymerization reaction can take place according to several methods. The cationic polymerization is one of these methods; it's a type of ionic polymerization (anionic and cationic). It takes place by cycle or double bonds opening. The most important step in the propagation reaction is the breakdown of the carbenium bond. Lewis acids such as BF_3 can be used as initiators in such reactions (Goethals and Penczek 1989; Wool et al. 2000; Olson 2009).

Acrylic monomers can be copolymerized with a low molecular weight molecules such as styrene or vinyl acetate to obtain the resins (Erbil 2000;La Scala and Woo 2005). The copolymerization is used for making a polymeric product with specifically desired properties.

In the present chapter the copolymerization reaction was carried out on a laboratory scale in a reactor with stirring and heating. The acrylatedepoxidized sunflower oil (AESFO) was mixed with styrene which acts as a solvent for the resin and at the same time as a co-monomer for improving the mechanical properties of the finished product. The initiator used is $BF_3$and the cobalt (Co) was also used as a catalyst to accelerate the curing of the finished product. First, the AESFO was mixed well with styrene, and then BF_3 and Co. were added. The casting technique was used to obtain platelets. For this purpose, the synthetic product is cast in silicone molds. The drying is carried out at 60°C in an oven for 12 hours.

The concentrations of AESFO, styrene, BF_3 and Co were varied in order to optimize the AESFO/styrene/BF_3/Co ratio of the reaction. The prepared samples were evaluated according to their appearance and then characterized in terms of tensile properties (stress at break, Young's modulus and elongation at break) (Tables 1, 2 and 3). From Tables 1 and 2, the amount of cobalt giving the best tensile properties is 0.02% for both concentrations of BF_3 (0.5 and 1%). However, the formulation with 0.5% BF_3 exhibited better stress at break, Young's modulus and elongation at break compared to the formulation with 1% BF_3.

Table 1. Tensile properties of formulation based on AESFO and styrene with 0.5% BF_3

Co%	Stress at break σ (MPa)	Young's modulus E (MPa)	Elongation at break (%)
0%	1.95 ± 0.19	04.40 ± 0.54	73.00 ± 3.65
0.01%	2.18 ± 0.08	15.92 ± 1.06	38.76 ± 0.83
0.02%	4.40 ± 0.15	19.52 ± 0.73	47.03 ± 0.82
0.03%	1.22 ± 0.05	05.77 ± 0.49	31.20 ± 0.87

Table 2. Tensile properties of formulation based on AESFO and styrene with 1% BF_3

Co%	Stress at break σ (MPa)	Young's modulus E (MPa)	Elongation at break (%)
0%	1.42 ± 0.15	2.46 ± 0.79	78.77 ± 10.54
0.01%	1.99 ± 0.09	4.40 ± 0.24	26.35 ± 2.5
0.02%	2.25 ± 0.19	15.68 ± 0.92	37.71 ± 3.06
0.03%	1.22 ± 0.10	6.51 ± 0.3	23.94 ± 0.59

After fixed the amounts of Co and BF_3, respectively at 0.02 and 0.5% were chosen in order to optimize styrene. For that, 30, 40 and 50% of styrene was mixed to AESFO, BF_3 and Co. Stress-strain results of styrene optimization at the previous fixed amounts of Co and BF_3 are presented in Table 1.3. These results show that there is a sharp increase in stress at break and Young's modulus at 50% styrene that have reached respectively 4.40 MPa and 19.52 MPa. For that the best formulation of the resine is AESFO/

styrene/BF_3/Co was found at 50% of AESFO, 50% of styrene, 0,5% of BF_3 and 0,02% of Co.

Table 3. Tensile properties of formulation on AESFO, BF_3 and Co with 30, 40 and 50% of styrene

Styrene%	Stress at break σ (MPa)	Young'smodulus E (MPa)	Elongation at break (%)
30%	1.22 ± 0.09	5.27 ± 0.29	35.60 ± 0.75
40%	1.37 ± 0.11	5.08 ± 0.27	39.76 ± 0.94
50%	4.40 ± 0.15	19.52 ± 0.73	47.03 0.82

2.2. Reinforcement Preparation

Alfa fibers were washed with water (2% detergent) to remove contaminants and dust. They were dried at room temperature and cut into 1 cm length. Then the fibers were crushed with a blade crusher to obtain length ranging from 0.4 to 1cm. Finally the fibers were dried in vacuum oven 60°C. The fibers obtained were immersed in a 5% NaOH solution for 6 hours at room temperature with continuous stirring. After that, they were rinsed with fresh water to remove the excess of NaOH, then with diluted acetic acid in order to neutralize their pH. Finally they were rinsed one last time with distilled water and dried for 12h at 60°C. These fibers will be designated as treated alfa fibers (TAF). After being sifted, an interval between "250-1000μm" was taken.

In order to well understand the alkali treatment effect on the alfa fibers, a scanning electron microscopy (SEM) test was done. This test is generally employed for the extensive morphological inspection and interfacial characterization.

The SEM photographies of both treated and untreated fibers were carried out (Figure 3).

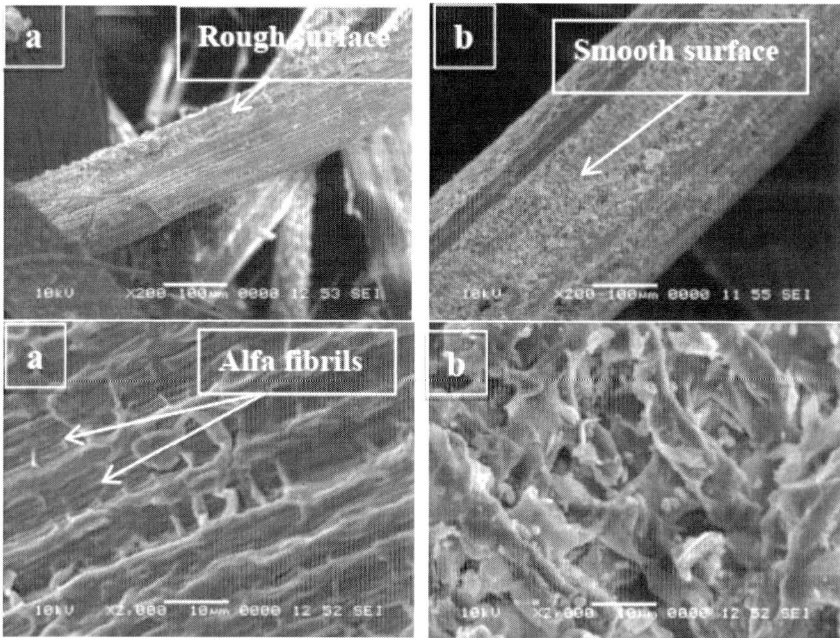

Figure 3. SEM photography: (a) treated alfa fibers; (b) untreated alfa fibers.

The comparison between the micrographs of treated and untreated alfa fibers shows morphological changes in the surface. The treated alfa surface (Figure 1a) is rough with presence of fibrils on the surface, while the untreated fiber surface (Figure 3b) is smooth and the fibrils were completely covered. This can be explained by the removal of non-cellulosic materials (lignin, hemicellulose and wax) by the alkali treatment which increases the interfacial adhesion between the fiber and the matrix (Faruk et al. 2012; Borchani et al. 2015; Kadem et al. 2018).

2.3. Biocomposites Preparation and Characterization

The preparation of the biocomposites was made by mixing the optimized resin with the treated alfa fiber (TAF) at different percentages (5%, 7,5% and 10%) and with 5% untreated alfa fibers (UAF) to see the effect of the alkali treatment on adhesion between the fibers and the resin. The resulting solution was poured into silicone molds which were placed in the oven at 60°C for 12 h.

2.3.1. Tensile Tests

To optimize the best percentage, the tensile properties results of the prepared biocompositesare showed in Table 4.

From Table 4, it can be noticed that the tensile properties of treated alfa fibers are better than those of the untreated ones and the results also shows that the biocomposite with 7.5% TAF presented the highest stress at break (5.12 MPa) and the highest Young's modulus (73.67 MPa), However, the elongation at break showed a considerable decrease with the addition of the alfa fibers. For that the biocomposite with 7.5% of treated alfa fibers was selected for the rest of the tests.

Table 4. Tensile properties of biocomposites with 5% UAF and 5, 7.5 and 10% TAF

Biocomposite	Stress at break σ (MPa)	Young's Modulus E (MPa)	Elongation at break (%)
Biocomposite (5% UAF)	1.96 ± 0.25	15.17 ± 0.52	20.95 ± 0.61
Biocomposite (5% TAF)	3.11 ± 0.12	19.30 ± 0.96	28.28 ± 3.23
Biocomposite (7.5% TAF)	5.12 ± 0.39	73.67 ± 0.41	27.49 ± 1.07
Biocomposite (10% TAF)	3.34 ± 0.15	23.29 ± 0.39	23.93 ± 1.41

2.3.2. Scanning Electron Microscopy

The scanning electron micrographs of biocomposites for both treated and untreated fibers are illustrated in Figure 4.

Figure 4a shows a rigid interfacial adhesion between the treated alfa fiber and the resin. Due to the alkaline treatment, this bonding is due to the rough surface of alfa fibers which causes a better interlocking between resin and fibers and to the chemical interactions due to the hydrogen bonding between the hydroxyl groups of cellulose in the fibers and the carboxylic groups in the resin making the biocompositemore rigid. On the other hand,

Figure 4b shows a poor adhesion between the untreated fibers and the resin. From the above results, the alkali treatment improved the adhesion at the fibers - resin interface (Williams and Wool 2000; Mechakra et al. 2015;Kadem et al. 2018).

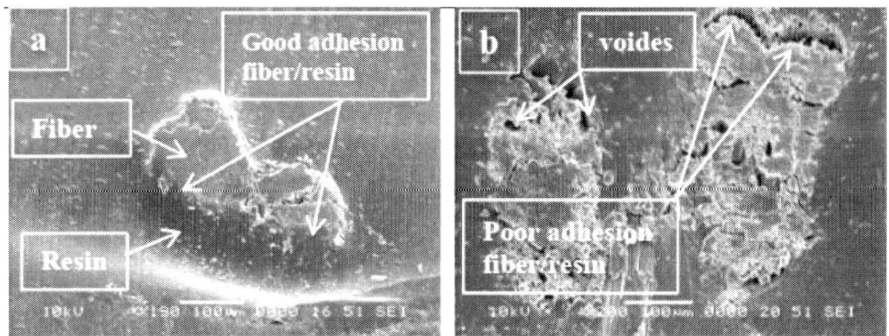

Figure 4. SEM photography of biocomposite (cross-sectional views): (a) biocomposite with treated fibers; (b) biocomposite with untreated fibers.

2.3.3. Thermogravimetric Analysis

To evaluate the influence of the added alfafibes on the thermal stability of the biocomposite a thermogravimetric analysis (TGA) was performed (Figure 5). The corresponding data are given in Table 5.

Table 5. Thermogravimetric data of the selected resin and biocomposite

	Stage I		Stage II		Stage III		
Sampels	T_{max}	W1	T_{max}	W2	T_{max}	W3	W_{total}
Resin	-	-	-	-	416	92	92
Biocomposite	240	4	344	12	423	74	90

W: Weight loss (%).

T_{max}: Temperature of maximum weight loss (°C).

W_{total}: Total weight loss (%).

The thermal decomposition (Figure 5) involved at least three steps of degradation for the selected biocomposite (with 7.5% TAF) and one step for the selected resin.

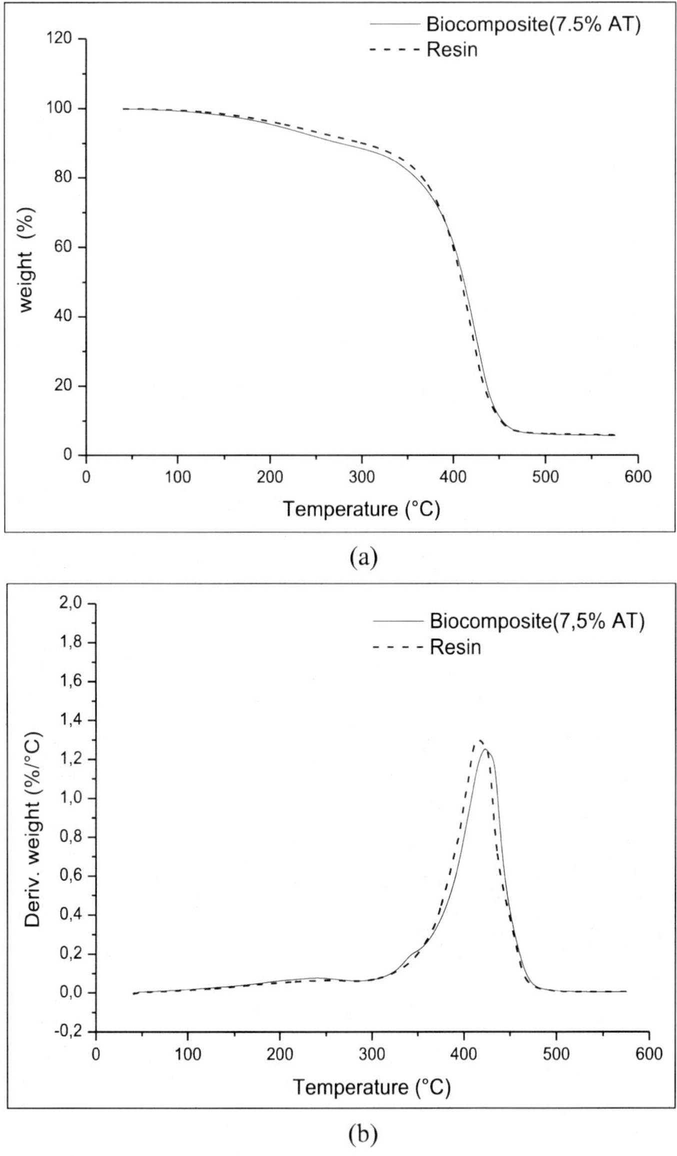

Figure 5. Thermogravimetric curves of selected resin and biocomposite: (a) weight loss; (b) derivative weight loss.

For the biocomposite, the first step (200-287°C) corresponds to the degradation of alfa fibers (hemicelluloce and lignine), the second step (287-357°C) is associated to the thermal decomposition of cellulose in alfafibres and the third step (357-471°C) is the same for the resin and corresponds to the depolymerisation of the styrene monomer, after that the weight loss varies slightly (Suzuki and Wilkie 1995; Islam et al. 2011; Boubekeur et al. 2015).

A better thermal stability is observed in the biocomposite (T_{max} = 424°C) compared to the resin (T_{max}= 416°C) and the total weight loss decreased in the case of the biocomposite compared to the neat resin which can be explained by the fact that the incorporation of treated alfa fibers exerted a stabilizing effect on the thermal degradation in the biocomposite which is due to the hydrogen bonding between the hydroxyl groups of cellulose and the carboxylic groups in the resin.

2.3.4. Water Absorption Test

Water absorption test was carried out on the selected resin and biocomposite. From the results shown in Figure 6, it is observed that there is a water absorption in both samples.

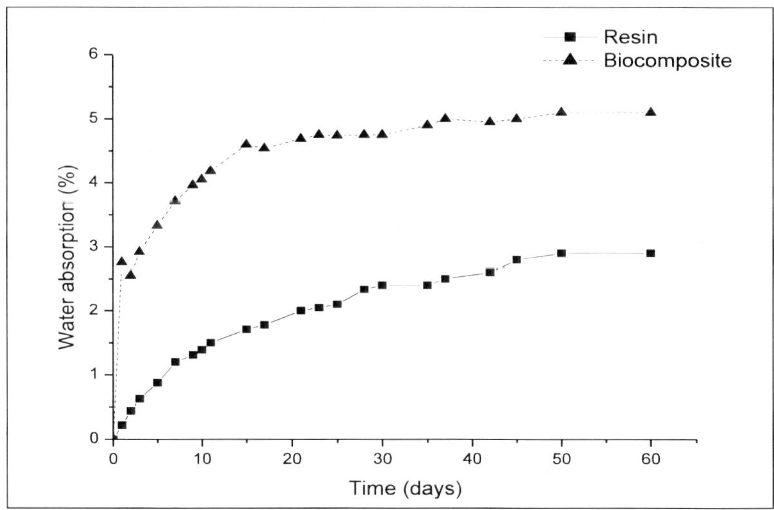

Figure 6. Water absorption of selected resin and biocomposite.

It occurred in three steps: the first step shows a rapid absorption during the first fifteen days, after a slow absorption step between the fifteenth and the twenty fifth day, followed by a saturation step. After sixty days of immersion the rate of water absorption at the end of the test is 2.8% in the resin and 5.1% in the biocomposite.

These results show that the absorption rate in the biocomposite is higher than in the resin. This can be explained by the capillary effect of natural fibers, this effect is due to the hydrophilic property caused by the hydroxyl groups in cellulose which is the major component of these natural fibers and this in accordance with the literature concerning natural fibers such as jute and sisal fiber (Espert et al. 2004; Yang et al. 2011).

3. Biodegradation Study of Biocomposites

In order to study the biodegradability of these new biocomposites, a respirometry test was carried out in the soil as well as a burial test.

3.1. Respirometry Test

The respirometry test is based on measuring the respiration of microorganisms degrading the test material. More specifically, the percentage of biodegradation of the test material is determined from the level of O_2 consumed or CO_2 released. This measured rate is then related to the maximum theoretical amount of oxygen or carbon dioxide that could have been consumed or produced by the microorganisms. The initial carbon content of the material must be known to measure the percentage of biodegradation (Bastioli 2005) [29].

In this study, the respirometry test was carried out according to the ISO 14855-2 standard. It is dedicated to the measurement of the amount of carbon dioxide produced by the microorganisms during the duration of the test under well controlled conditions (humidity pH and C/N ratio).

The test was carried out on the selected resin and biocomposite during six months in the laboratory. The samples were doubled for accuracy. A mass of each sample containing approximately 500 mg of carbon was taken according to the standard ISO 14855-2. The samples were cut into small pieces and mixed with 200 g of soil already characterized and analyzed and contains water at 80% of its retention capacity. These different mixtures were introduced into 500 ml Erlenmeyer flasks.

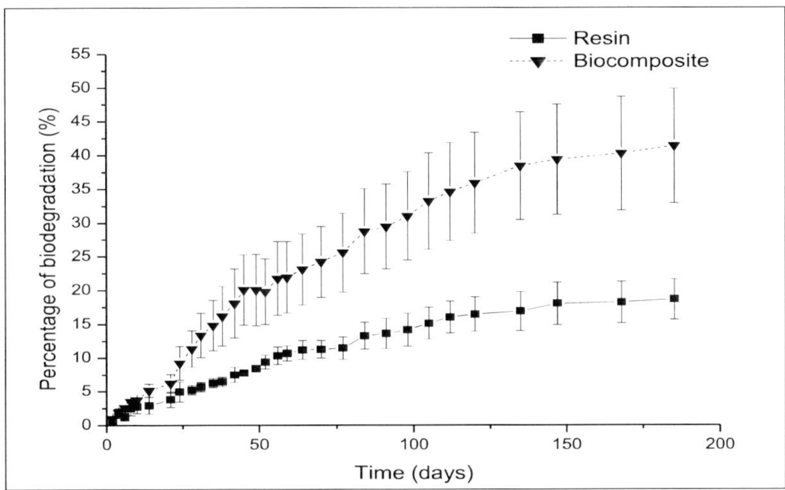

Figure 7. Percentage of biodegradation of selected resin and biocomposite.

Two blank samples were added as control; they contain only the soil and were subjected to the same conditions as the test material.

The percentage of biodegradation of the sample was calculated from the CO_2 levels released (NI 2007) according to equation (1).

$$\text{Dt}\,\% = \frac{\sum(\text{CO}_2)\text{t} - \sum(\text{CO}_2)b}{Th\,CO_2}\,X\,100 \tag{1}$$

where $\sum(CO_2)t$ is the cumulative amount of carbon dioxide, in grams, evolved in the test vessel between the start of the test and time t;

$\sum(CO_2)b$ is the mean cumulative amount of carbon dioxide, in grams, evolved in the blank vessels between the start of the test and time (take the mean of the values obtained for the two blanks);

$ThCO_2$ is the theoretical amount of carbon dioxide, in grams, evolved by the test material.

The results are shown in Figure 7. The biodegradation took place in three phases:

- The first phase: for eight days there is a low release of CO_2. This is a latency phase corresponding to the stage of microorganism's adaptation.
- The second phase: between the eighth to the hundred and thirty-fifth day, it's the phase of biodegradation; a large amount of CO_2 has been produced. This phase corresponds to the hydrolysis of the materials most sensitive to biodegradation by microorganisms such as alfa fiber which contains the hydrolysable groups (cellulose) and the shortest chains in the resin synthesized.
- The third phase: it is the stationary phase. It starts from the one hundred and thirty-fifth day where a plateau is obtained. It means that the cumulative amount of CO_2 released stabilizes, and therefore, there is a decrease of the microbial activity that corresponds to the biodegradation of the most resistant substances that are the long chains of the synthesized resin.

At the end of the respirometry test, the rate of degradation is 19% in resin and 41% in biocomposite.

3.2. Burial Test

The biodegradation of the synthesized materials has been studied in the soil. The test was carried out on the selected resin and biocomposite during one year in the laboratory. The samples were buried in a glass container that contains soil already characterized and analyzed.

These samples (selected resin and biocomposite) were cut into small squares on the order of 1 cm^2 and weighed to obtain the initial mass (m_0). The samples were doubled for accuracy and they were buried in the soil at a depth of 2-3 cm. The biodegradation was measured by the mass loss of samples.

The mass loss is calculated according to equation (2).

$$\Delta m\ \% = \frac{m_t - m_0}{m_0} X\ 100 \tag{2}$$

where:

Δm: loss of mass in grams.

m_0: initial mass before burial in the soil.

m_t: mass taken at time t.

The evolution of the mass loss of buried materials is given in Figure 8. Mass loss was observed from the first days of test. After one year of burial, the loss of mass was 7.8% for the resin and 10.2% for the biocomposite.

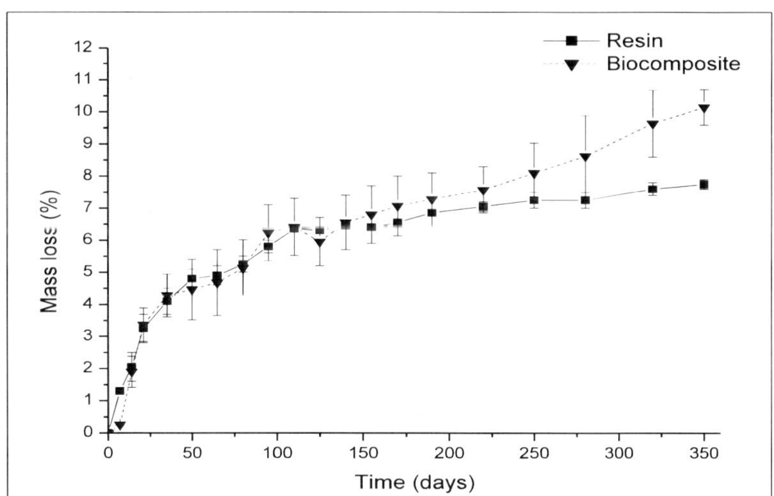

Figure 8. Mass loss of selected resin and biocomposite.

According to the results of biodegradation found in respirometry and burial tests, the synthesized materials are biodegradable and this biodegradation is more important in biocomposite, this is due to the incorporation of alfa fibers which are naturally biodegradable.

Conclusion

The elaboration of new biocomposites based on sunflower oil and alfa fibers as renewable materials were successfully made out. The resin of this biocomposite was synthesized after chemical modification and copolymerization of sunflower oil then alfa fibers were added as reinforcement to the resin after alkali treatement, this treatement exhibited better interfacial adhesion between the fiber and the resin and improved the mechanical and the thermal properties.

Theprepared biocomposites have good mechanical and thermal properties. This expands their areas of application. These materials can be used as protective coatings for metals, building materials, insulation materials and in interior decoration.

The biodegradation tests (respirometry and burial in the soil) showed a high capacity of biodegradation in a relatively short time compared to other composites from petroleum-based source.

References

Bastioli, C. (2005). *Handbook of biodegradable polymers*. Smithers Rapra Publishing.

Benaniba, M. T., Belhaneche-Bensemra, N. and Gelbard, G. (2007). Kinetics of tungsten-catalyzed sunflower oil epoxidation studied by 1H NMR. *European Journal of Lipid Science and Technology*, 109(12): 1186-1193.

Benaniba, M., Belhaneche-Bensemra, N. and Gelbard, G. (2008). Epoxidation of sunflower oil with peroxoacetic acid in presence of ion exchange resin by various processes. *Energy Education Science and Technology*, 21(1-2): 71-82.

Borchani, K., Carrot, C. and Jaziri, M. (2015). Untreated and alkali treated fibers from Alfa stem: effect of alkali treatment on structural, morphological and thermal features. *Cellulose*, 22(3): 1577-1589.

Boubekeur, B., Belhaneche-Bensemra, N. and Massardier, V. (2015). Valorization of waste jute fibers in developing low-density polyethylene/poly lactic acid bio-based composites. *Journal of Reinforced Plastics and Composites*, 34(8): 649-661.

Carbonell-Verdu, A. et al. (2015) Development of environmentally friendly composite matrices from epoxidized cottonseed oil. *European Polymer Journal,* 63(0): 1-10.

Crivello, J. and Narayan, R. (1992). Epoxidized triglycerides as renewable monomers in photoinitiated cationic polymerization. *Chemistry of materials*, 4(3): 692-699.

Erbil, Y. H. (2000). *Vinyl acetate emulsion polymerization and copolymerization with acrylic monomers*. CRC press.

Espert, A., Vilaplana, F. and Karlsson, S. (2004). Comparison of water absorption in natural cellulosic fibres from wood and one-year crops in polypropylene composites and its influence on their mechanical properties. *Composites Part A: Applied Science and Manufacturing*, 35(11): 1267-1276.

Faruk, O. et al. (2012). Biocomposites reinforced with natural fibers: 2000–2010. *Progress in Polymer Science*, 37(11): 1552-1596.

Goethals, E. and Penczek, S. (1989). Cationic ring-opening polymerization: introduction and general aspects. *Pergamon Press plc, Comprehensive Polymer Science: the Synthesis, Characterization,Reactions & Applications of Polymers*, 3: 711-717.

Irinislimane, R. and Belhaneche-Bensemra, N. (2012). Application of fourier transform infrared (FT-IR) spectroscopy to the study of the modification of epoxidized sunflower oil by acrylation. *Applied Spectroscop,* 66(12): 1420-1425.

Islam, M. S., Pickering, K. L. and Foreman, N. J. (2011). Influence of alkali fiber treatment and fiber processing on the mechanical properties of hemp/epoxy composites. *Journal of Applied Polymer Science,* 119(6): 3696-3707.

Kadem, S., Irinislimane, R. and Belhaneche-Bensemra, N. (2018). Novel Biocomposites Based on Sunflower Oil and Alfa Fibers as Renewable Resources. *Journal of Polymers and the Environment*, 26(7): 3086-3096.

Kalnin'sh, K. (2010). Excited biradical states in oxirane ring opening. *Russian Journal of Applied Chemistry*, 83(5): 858-863.

Kaplan, D. L. (1998). *Introduction to biopolymers from renewable resources, in Biopolymers from renewable resources*. Springer: 1-29.

Kaya, Y., Jocic, S. and Miladinovic, D. (2012) *Sunflower, in Technological Innovations in Major World Oil Crops*, Volume 1. Springer: 85-129.

La Scala, J. and Wool, R. P. (2005). Property analysis of triglyceride-based thermosets. *Polymer,* 46(1): 61-69.

Malacea, R. and Dixneuf, P. H. (2010). *Alkene Metathesis and Renewable Materials: Selective Transformations of Plant Oils, in Green Metathesis Chemistry*. Springer. 185-206.

Mechakra, H. et al. (2015). Mechanical characterizations of composite material with short Alfa fibers reinforcement. *Composite Structures,* 124(0): 152-162.

Norme Internationale (NI) ISO/FDIS 14855-2, (2007). *Determination of the ultimate aerobic biodegradability of plastic materials under controlled composting*.

Olson, C. (2009). *Varnish forming properties of sunflower oil and how they relate to its use as fuel in diesel tractors*.

Pfister, D. P. and Larock, R. C. (2010). Thermophysical properties of conjugated soybean oil/corn stover biocomposites. *Bioresource Technology*, 101(15): 6200-6206.

Sánchez-Muniz, F. J. and Cuesta, C. (2003). Sunflower Oil, in *Encyclopedia of Food Sciences and Nutrition* (Second Edition), B. Caballero, Editor. Academic Press: Oxford. 5672-5680.

Seniha Güner, F., Yağcı, Y. and Tuncer, A. (2006). Erciyes, Polymers from triglyceride oils. *Progress in Polymer Science*, 31(7): 633-670.

Suzuki, M. and Wilkie, C. A. (1995). The thermal degradation of acrylonitrilebutadiene-styrene terploymer grafted with methacrylic acid. *Polymer Degradation and Stability*, 47(2): 223-228.

Vollmann, J. and Rajcan, I. (2009). *Oil crops*. Vol. 4. Springer Science & Business Media.

Williams, G. I. and Wool, R. P. (2000). Composites from natural fibers and soy oil resins. *Applied Composite Materials*, 7(5-6): 421-432.

Wool, R. et al. (2000). High modulus polymers and composites from plant oils. *Google Patents*.

Yang, Y. et al. (2011). Mechanical property and hydrothermal aging of injection molded jute/polypropylene composites. *Journal of Materials Science*, 46(8): 2678-2684.

About the Editor

Érico de Sá Petit Lobão

Researcher collaborator DCR (FAPEMAT/CNPq) - Embrapa Agrossilvipastoril (Feb / 2018 - Apr/2019). Project: "Farm-Livestock-Forest Integration for breeding and rearing systems". Supervisors: DSc Bruno Carneiro e Pedreira; DSc Luciano Bastos Lopes. Experiences: bases of integration between crop-livestock-forest (iLPF); animal production in iLPF; efficiency and reproductive management, protocols for induction of precocity, physiology of reproduction; parasitic control (endo and ecto); precision animal husbandry (management of rangeland, pasture management); bromatological evaluation with NIR spectrophotometer; accomplishment of technical event; postgraduate teaching activity; field routine and research. Substitute Professor - CEUNES / UFES (May-Aug /

2017). Experiences: teaching activity in undergraduate; production of ruminants; production of non-ruminants; physiology and animal anatomy; environmental management and management; directed studies in silvopastoral systems; academic routine. Pronatec Professor - IFBA / Campus-Ilhéus (2014-2015). Experiences: teaching activity in technical education; methods of preserving food of animal and plant origin; animal husbandry; course preparation; academic routine. Post-doctoral in Animal Science in the Tropics - UFBA (2013-2014). Supervisor: DSc Gleidison Giordano P. Carvalho. Project: "Peanut and cotton pie, from the production of biodiesel, in diets for sheep and goats". Experiences: Postgraduate teaching activity; conducting events, reviewing papers and academic-scientific articles; management and academic routine. Doctor of Zootechnics - Forragicultura - UFPB (2009-2013). Advisor: DSc Albericio Pereira Andrade. TESE: "The effect of organic fertilization on the production and quality of sunflower cake". Experiences: forage farming; animal nutrition; evaluation and production of dryland food for the production of ruminants; Teaching activity at undergraduate level; conducting academic events; instructor in mini courses; presentation and orientation of academic-scientific works; composition of examining bank. Master in Applied Zoology - Wild Animal Breeding - UESC (2004-2006). Advisor: DSc Sérgio Luiz G. Nogueira Filho. Dissertation: "Analysis of the conflicts between rural producers and wild mammals in the cacao region of Southern Bahia - central corridor of the Atlantic Forest". Experiences: capture, containment, breeding and nutritional, sanitary and reproductive management of wild animals; ethology; monitoring and traceability; survey and resolution of fauna-man conflicts; teaching in animal welfare, ecology and wild breeding; the accomplishment of academic-scientific events; teaching activity at graduation. Passant - University of the Republic - URU (May-June 2000). Improvement in Ruminant Production (200h). Tutors: DSc Steban Krall and DSc Diego Antonio Mattiauda. Experiences: management of facilities and equipment; animal nutrition; ethology; dairy cattle; beef cattle; production systems; research and rural extension; agroindustry of meat. Bachelor in Animal Science - UFV (1996-2001). TCC: Seminar "Sistema Cabruca". Project: "Recommendation of Fiber

Levels and their Effectiveness in Industry By-Products in Diets for Dairy Goats". Internships: Poultry (1998); Caprinocultura-UFV (1999-2001). Experiences: sanitary, nutritional and reproductive management of birds and small ruminants; animal nutrition; ethology; lab routine and capril; production in confinement. Other: Coordinator of International Network for Silvopastoralism (INES; 2015-current); Deputy Leader - GPCP / UESC (2012-current); Scientific Director - FUNPAB (2014-current); Adviser - SRI (2002-2014); Ad hoc consultant - CDAC (2005-2012) - Experiences: conservation-productive; use-multiple and rational use of natural resources; cabruca system; agroforestry systems.

INDEX

D

E

F

N

O

P

R

S

T

U

V

W

Y

Related Nova Publications

Germination: Types, Process and Effects

Editors: Rosalva Mora-Escobedo, PhD, Cristina Martinez, and Rosalía Reynoso

Series: Plant Science Research and Practices

Book Description: *Germination: Types, Process and Effects* is a book that brings together the contribution of new and relevant information from many experts in the fields of food and biological sciences, nutrition, and food engineering, to provide the reader with the latest information of fundamental and applied research in the role of edible seeds and discuss the benefits of consuming them.

Hardcover ISBN: 978-1-53615-973-8
Retail Price: $230

Phytochemicals: Plant Sources and Potential Health Benefits

Editor: Iman Ryan

Series: Plant Science Research and Practices

Book Description: The opening chapter of *Phytochemicals: Plant Sources and Potential Health Benefits* discusses macronutrients and micronutrients from plants along with their benefits to human health.

Hardcover ISBN: 978-1-53615-478-8
Retail Price: $230

To see a complete list of Nova publications, please visit our website at www.novapublishers.com

Related Nova Publications

Plant Dormancy: Mechanisms, Causes and Effects

Editor: Renato V. Botelho

Series: Plant Science Research and Practices

Book Description: Dormancy is a mechanism found in several plant species developed through evolution, which allows plants to survive in adverse conditions and ensure their perpetuation.

Hardcover ISBN: 978-1-53615-380-4
Retail Price: $160

Micropropagation: Methods and Effects

Editor: Valdir M. Stefenon, Ph.D.

Series: Plant Science Research and Practices

Book Description: Plant micropropagation is one of the most classical and widespread biotechnological tools used around the world. Undoubtedly, this technique brought quite important advances to our knowledge about morphological, physiological and developmental patterns of plants, to the progress of genetic breeding and to the establishment of the genetic engineering, among others.

Softcover ISBN: 978-1-53614-968-5
Retail Price: $82

To see a complete list of Nova publications, please visit our website at www.novapublishers.com